Forest Landscape Ecology
Transferring Knowledge to Practice

Forest Landscape Ecology
Transferring Knowledge to Practice

Edited by

Ajith H. Perera
Ontario Forest Research Institute
Sault Ste. Marie, Ontario
Canada

Lisa J. Buse
Ontario Forest Research Institute
Sault Ste. Marie, Ontario
Canada

and

Thomas R. Crow
USDA Forest Service
Arlington, Virginia
USA

Ajith H. Perera
Ontario Forest Research
  Institute
1235 Queen Street East
Sault Ste. Marie
Ontario P6A 2E5
Canada
E-mail:
ajith.perera@mnr.gov.on.ca

Lisa J. Buse
Ontario Forest Research
  Institute
1235 Queen Street East
Sault Ste. Marie
Ontario P6A 2E5
Canada
E-mail:
lisa.buse@mnr.gov.on.ca

Thomas R. Crow
USDA Forest Service
WFWAR
1601 N. Kent Street
Arlington, VA 22209
USA
E-mail: tcrow@fs.fed.us

Library of Congress Control Number: 2006925854

ISBN-10: 0-387-34243-5   eISBN-10: 0-387-34280-X
ISBN-13: 978-0-387-34243-6   eISBN-13: 978-0-387-34280-1

Printed on acid-free paper.

© 2007 Springer Science+Business Media, LLC
All rights reserved. This work may not be translated or copied in whole or in part without the written permission of the publisher (Springer Science+Business Media, LLC, 233 Spring Street, New York, NY 10013, USA), except for brief excerpts in connection with reviews or scholarly analysis. Use in connection with any form of information storage and retrieval, electronic adaptation, computer software, or by similar or dissimilar methodology now known or hereafter developed is forbidden.
The use in this publication of trade names, trademarks, service marks, and similar terms, even if they are not identified as such, is not to be taken as an expression of opinion as to whether or not they are subject to proprietary rights.

9 8 7 6 5 4 3 2 1

springer.com

Foreword

Forested landscapes have provided many important testing grounds for the development and application of landscape ecological principles and methods in North America. This central role of forests in landscape ecology emerged for several reasons. Forest cover is prominent in many regions of North America, from the temperate deciduous forests of the east to the coniferous forests of the north and west. Changes in forest spatial patterns are readily apparent to the human eye—natural disturbances and timber harvests alter the arrangement of forest age classes across the landscape and this, in turn, influences many species and ecosystem processes; land-use changes have produced profound fluctuations in forest cover over several centuries; increasing residential development in rural areas is often concentrated within forests; and public lands include many forested landscapes. Management actions, such as varying the amount, size, and location of harvests, also represent landscape-scale "experiments" that provide valuable opportunities for study. Finally, forest patterns are readily detectable from remote imagery, and are thus amenable to study at broad scales. For these reasons, forests have provided motivation and many opportunities for studying the complex relationships between patterns and processes in many areas.

The importance of landscape-level considerations in the management and conservation of forested landscapes has become increasingly important, and a variety of stakeholders are involved. The discipline of landscape ecology has developed concepts and methods that can be directly applied in forested landscapes, but to be most useful, these need to be more widely available. Included are principles and theory that relate spatial patterns at multiple spatial and temporal scales to ecological and anthropogenic drivers; methods for quantifying and evaluating spatial patterns in both discrete and continuous variables; and models for projecting the consequences of alternative scenarios. This book contains numerous examples from landscape ecology and concrete suggestions for increasing its utility in forest ecology and management.

Increasing the applicability of landscape ecological concepts and methods in the management of forested landscapes is a worthy goal. Much remains to be done, although there is perhaps a longer and richer history of such activities than suggested by the new terminology of "knowledge transfer." This term emerged relatively recently as a more inclusive term for integrating the basic and applied aspects of science, and providing opportunities for practitioners to learn about recent scientific developments. Although basic and applied sciences have been well integrated in landscape ecology as it matured, "bridging the gap" seems to be a perennial challenge. Thus, acknowledging the importance of knowledge transfer and continually improving its effectiveness remains critical. Scientists and practitioners need a two-way dialogue in which the science is made clear, accessible, and relevant to those seeking to apply it. In turn, the practitioners must make their management needs and challenges clear, as these often catalyze new developments in the science. It is also of paramount importance that the researchers who are producing new knowledge work actively to transfer that knowledge to other users. Communicating with users should be an integral part of the overall process, and researchers should seek and use all available communication opportunities.

By providing interesting examples and a synthesis of knowledge transfer, this volume makes a significant contribution to the applications of landscape ecology in forested landscapes. However, as the authors note, knowledge transfer is necessary but not sufficient for applications of landscape ecology to be successful. The chapters within provide readers with ideas and examples for successfully translating the science of landscape ecology into practice. The book should be of broad interest to all those interested in understanding and managing forested landscapes, and in particular to current and future researchers committed to making the knowledge they develop available to a wide audience of users. This volume clearly helps chart the course.

Monica G. Turner
Eugene P. Odum Professor of Ecology
University of Wisconsin
Madison, Wisconsin
March 2006

Preface

Forest landscape ecology has matured rapidly over the past two decades in North America, and the result has been the development of a substantial body of published knowledge. From our vantage point as landscape ecologists in forest management agencies, we have seen the potential for using landscape ecological knowledge in forest policy development, land-use planning, and resource management increase over the same period. We have also observed the difficulties faced by forestry professionals in their attempts to apply landscape ecological concepts. We see a growing role for those who develop landscape ecological knowledge: to synthesize and transfer that knowledge to users to ensure appropriate application of this knowledge. At the same time, transfer and extension are relatively new concepts to many researchers. It is this context that inspired us to compile this book.

Our goal is to introduce the topic of knowledge transfer to researchers in forest landscape ecology and to demonstrate how transfer efforts can be effective. We do so by reviewing general aspects of knowledge transfer and extension, critically examining aspects of transfer that are unique to forest landscape ecology, and highlighting several successful examples of knowledge transfer. This book captures the knowledge, experience, and insights of a group of authors with diverse backgrounds, ranging from university academics to researchers in forest management agencies and nongovernmental organizations, and diverse expertise, ranging from extension and knowledge transfer to landscape modeling. Our intended primary readership is developers of forest landscape ecological knowledge, whether they are academics, researchers, technologists, or graduate students.

This book is not a comprehensive treatise on knowledge transfer; it is meant to be a primer for landscape ecologists, written primarily by landscape ecologists. We encourage readers to consult the vast body of literature on organizational learning, extension, and knowledge transfer to gain an in-depth appreciation of these subjects. As well, we were unable, despite our best efforts, to find a user willing to share their perspective in this book. This does not diminish the importance of the need for readers to understand the user's expectations.

We thank the colleagues who improved the chapter contents by peer review: Jim Baker, Larry Biles, Jiquan Chen, Joe Churcher, Tom Clark, David DeYoe, Michael Drescher, Dave Euler, Paul Hessburg, Louise Levy, Jim Manolis, Eric

Norland, Bruce Pond, Volker Radeloff, Janet Silbernagel, Susan Smith, Fred Swanson, Michael Wimberly, and Kim With. We also gratefully acknowledge those who assisted us in producing this book: Trudy Vaittinen for improving the illustrations, Janet Slobodien for being our patient liaison at Springer, and Geoff Hart for language editing.

Ajith H. Perera, Lisa J. Buse, and Thomas R. Crow
March 2006

Contents

1. **Knowledge Transfer in Forest Landscape Ecology: A Primer**   1
   Ajith H. Perera, Lisa J. Buse, and Thomas R. Crow

2. **Transfer and Extension of Forest Landscape Ecology: A Matter of Models and Scale**   19
   Anthony W. King and Ajith H. Perera

3. **A Collaborative, Iterative Approach to Transferring Modeling Technology to Land Managers**   43
   Eric J. Gustafson, Brian R. Sturtevant, and Andrew Fall

4. **Development and Transfer of Spatial Tools Based on Landscape Ecological Principles: Supporting Public Participation in Forest Restoration Planning in the Southwestern United States**   65
   Haydee M. Hampton, Ethan N. Aumack, John W. Prather, Brett G. Dickson, Yaguang Xu, and Thomas D. Sisk

5. **Transferring Landscape Ecological Knowledge in a Multipartner Landscape: The Border Lakes Region of Minnesota and Ontario**   97
   David E. Lytle, Meredith W. Cornett, and Mary S. Harkness

6. **Applications of Forest Landscape Ecology and the Role of Knowledge Transfer in a Public Land Management Agency**   129
   Lisa J. Buse and Ajith H. Perera

7. **Moving to the Big Picture: Applying Knowledge from Landscape Ecology to Managing U.S. National Forests**   157
   Thomas R. Crow

8. **Fundamentals of Knowledge Transfer and Extension** 181
   A. Scott Reed and Viviane Simon-Brown

9. **Synthesis: What Are the Lessons for Landscape Ecologists?** 205
   Thomas R. Crow, Ajith H. Perera, and Lisa J. Buse

Index 215

Contributors

Ethan N. Aumack Grand Canyon Trust, 2601 N. Fort Valley Road, Flagstaff, AZ 86001, USA

Lisa J. Buse Ontario Ministry of Natural Resources, Ontario Forest Research Institute, 1235 Queen St. E., Sault Ste. Marie, ON P6A 2E5, Canada

Meredith W. Cornett The Nature Conservancy, 1101 West River Parkway, Suite 200, Minneapolis, MN 54415, USA

Thomas R. Crow USDA Forest Service, Research and Development, Environmental Sciences, 1601 N. Kent Street, Arlington, VA 22209, USA

Brett G. Dickson Colorado State University, Department of Fishery & Wildlife Biology, Fort Collins, CO 80521, USA

Andrew Fall Simon Fraser University, 8888 University Drive, Burnaby, BC V5A 1S6, Canada

Eric J. Gustafson USDA Forest Service, North Central Research Station, 5985 Highway K, Rhinelander, WI 54501, USA

Haydee M. Hampton Northern Arizona University, Center for Environmental Sciences and Education, NAU Box 5694, Flagstaff, AZ 86011, USA

Mary S. Harkness The Nature Conservancy, 1101 West River Parkway, Suite 200, Minneapolis, MN 54415, USA

Anthony W. King Oak Ridge National Laboratory, Environmental Sciences Division, Bldg 1509, MS 6335, Oak Ridge, TN 37831, USA

David E. Lytle The Nature Conservancy, 6375 Riverside Drive, Suite 50, Dublin, OH 43017, USA

Ajith H. Perera Ontario Ministry of Natural Resources, Ontario Forest Research Institute, 1235 Queen St. E., Sault Ste. Marie, ON P6A 2E5, Canada

John W. Prather* Northern Arizona University, Center for Environmental Sciences and Education, NAU Box 5694, Flagstaff, AZ 86011, USA

A. Scott Reed Oregon State University, College of Forestry Extension Service, Richardson Hall 109, Corvallis, OR 97331, USA

Viviane Simon-Brown Oregon State University, College of Forestry Extension Service, Richardson Hall 109, Corvallis, OR 97331, USA

Thomas D. Sisk Northern Arizona University, Center for Environmental Sciences and Education, NAU Box 5694, Flagstaff, AZ 86011, USA

Brian R. Sturtevant USDA Forest Service, North Central Research Station, 5985 Highway K, Rhinelander, WI 54501, USA

Yaguang Xu Northern Arizona University, Center for Environmental Sciences and Education, NAU Box 5694, Flagstaff, AZ 86011, USA

---

*Deceased February 2006

1

Knowledge Transfer in Forest Landscape Ecology: A Primer

Ajith H. Perera, Lisa J. Buse, and Thomas R. Crow

1.1. Why Should Forest Landscape Ecologists Focus on Knowledge Transfer?
1.2. What Factors Influence Knowledge Transfer?
    1.2.1. The Generation of Research Knowledge
    1.2.2. The Potential for Applications
    1.2.3. Users of the Knowledge
    1.2.4. Technological Infrastructure
    1.2.5. Barriers to Knowledge Transfer
1.3. What can Forest Landscape Ecologists Do to Advance Knowledge Transfer?
    1.3.1. Understand the Basics of Knowledge Transfer
    1.3.2. Play an Active Role
1.4. Summary
Literature Cited

1.1. WHY SHOULD FOREST LANDSCAPE ECOLOGISTS FOCUS ON KNOWLEDGE TRANSFER?

The science of landscape ecology has evolved rapidly from a relatively obscure topic, then a young discipline, to a popular focus for researchers. This evolution is reflected in a recent issue of *Ecology* (2005:86(8)) that is dedicated to the topic *landscape ecology comes of age*. As the knowledge base of landscape ecology expands and its range of topics broadens, researchers are becoming increasingly aware of the value of landscape ecology applications in managing both terrestrial and aquatic resources (Gutzwiller 2002; Liu and Taylor 2002).

AJITH H. PERERA and LISA J. BUSE • Ontario Ministry of Natural Resources, Ontario Forest Research Institute, 1235 Queen St. E., Sault Ste. Marie, ON P6A 2E5, Canada. THOMAS R. CROW • USDA Forest Service, Research and Development, Environmental Sciences, 1601 N. Kent Street, Arlington, VA 22209, USA.

In particular, the concepts of landscape ecology have increasingly been integrated into the study of forested environments in North America over the past two decades. In fact, the very first research paper in the inaugural issue of the journal *Landscape Ecology* addressed spatial patterns in a harvested forest landscape (Franklin and Forman 1987). The focus of forest landscape ecology, at least in a North American context, is large tracts of land where the cover is dominated by forests (i.e., the matrix) interspersed with areas where forest cover may be temporarily absent due to disturbances such as harvesting and fire (i.e., patches) (Perera and Euler 2000). This differs from the traditional milieu of landscape ecology, in which forest cover exists in patches (i.e., is fragmented) within a matrix of nonforested area and the transformation of forest patches to nonforest cover is usually permanent.

Viewing forested landscapes as broad-scale ecosystems and studying their composition, spatial patterns, spatial interactions, temporal change, and range of functions have direct applied value because most forests in North America are managed at broad scales to provide a range of uses: resource extraction, recreation, and conservation. Efforts to elucidate various broad-scale ecological patterns and processes in forested landscapes are essential to attaining the broad forest management goals of conserving forest biodiversity and attaining forest sustainability, as well as to understanding and mitigating the regional and global consequences of local forest management.

Although the value of landscape ecology applications is increasingly recognized, the transfer of knowledge in landscape ecology from those who develop it to those who apply it is not commonly identified as an explicit role for researchers. A literature search, for example, in the journals *Ecology*, *Ecological Applications*, *Forest Ecology and Management*, *Landscape Ecology*, *the Canadian Journal of Forest Research*, and *Forest Science* from 1960 to 2005 shows that no publications on landscape ecology or forest landscape ecology during that period contained any of the following keywords in the publications' titles, keywords, or abstracts: knowledge transfer, technology transfer, and extension. Furthermore, the topic of knowledge transfer was not addressed until 2004 at the annual meeting of the U.S. chapter of the International Association for Landscape Ecology, traditionally the principal gathering of landscape ecologists in North America. Although an extensive literature on knowledge transfer exists in social science journals, landscape ecologists do not readily encounter such studies. As a result, knowledge transfer remains for them an obscure topic of study.

Few developers of knowledge in forest landscape ecology, however, would dispute that the necessary next step in the evolution of the field is to move from the accumulating wealth of scientific and technical knowledge to applications of that knowledge. Forest landscape managers are in urgent need of such applications in formulating policies, planning the use and conservation of resources, and developing management strategies. As is the case with mature applied sciences such as agriculture and forestry, the progression from concepts and principles (i.e., knowledge in its primary form) to application of those concepts and principles requires forest landscape ecologists to engage explicitly and actively in knowledge transfer.

Our goal in this chapter is to introduce researchers and other knowledge developers in forest landscape ecology to the concept of knowledge transfer. To do so, we examine the key factors that influence knowledge transfer, focus on aspects that are unique to forest landscape ecology, and suggest a role for knowledge developers in the knowledge transfer process.

1.2. WHAT FACTORS INFLUENCE KNOWLEDGE TRANSFER?

First, let us define our terms. By *knowledge transfer* we mean a group of activities that increase the understanding of landscape ecology with the goal of encouraging application of this knowledge. *Technology transfer* implies a specific instance of knowledge transfer that increases levels of skill in the use of tools. *Extension* refers to a very broad group of practices geared toward knowledge and technology transfer that enable the successful application of knowledge. Although use of the term *extension* is common, we prefer the more specific term *knowledge transfer*, and we use that term in its broadest sense throughout this chapter, except when there is a specific need to differentiate between knowledge transfer and technology transfer.

We recognize five major factors that will influence knowledge transfer from the view of forest landscape ecology: the generation of research knowledge, the potential for application, the users of the knowledge, the infrastructure capacity, and the process by which knowledge is transferred. In the remainder of this section, we outline these factors and address how they interact during the process of knowledge transfer. We provide only a broad description since a detailed treatise on knowledge transfer principles and concepts is beyond the scope of this discussion. For that we refer the reader to other sources (e.g., Reed and Simon-Brown 2006; Rogers 1995).

1.2.1. The Generation of Research Knowledge

The increased academic interest in forest landscape ecology in North America is manifest in the growth of research capacity: almost all major universities have established graduate programs providing advanced training in this area of study. One indicator of increased activity is that 84 North American graduate thesis and dissertation titles contained the keyword "forest landscape" between 1990 and 2004, compared with only 5 prior to 1989. In addition, most major forest research agencies outside universities have developed directed research programs and projects on this topic. The resulting growth in the body of published scientific knowledge has been rapid, and is evident in the proliferation of research papers that specifically address forest landscape ecology (see Figure 1.1) and books in the field (e.g., Mladenoff and Baker 1999; Perera et al. 2000, 2004; Rochelle et al. 1999). All major journals that consider ecology and ecological applications now regularly publish research studies conducted on forested landscapes. The number of forest landscape ecology presentations delivered during scientific conferences, particularly by graduate students engaged in thesis research, has also increased considerably.

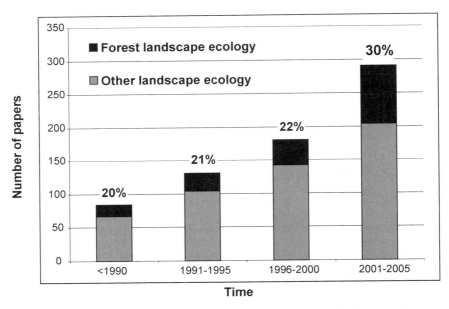

Figure 1.1. Research papers on forest landscape ecology and on other landscape ecology topics published in the journal *Landscape Ecology* from 1987 to 2005. (Percentages refer to the proportion of the total accounted for by papers on forest landscape ecology.)

The topics addressed by forest landscape ecology have also expanded and become increasingly specialized. Although early forest landscape ecology research focused primarily on habitat fragmentation and population dynamics, on the basis of island biogeography theory, recent research has embraced more of a systems view of forested landscapes, including attempts to apply other null models such as disturbance-resilience theory. For example, research papers published in the journal *Landscape Ecology* from 1987 to 2005 addressed a variety of aspects of forested landscapes, including the following: spatial heterogeneity (forest ecogeography, landscape indices, mapping and spatial pattern analyses of forest cover); forest landscape function (primary productivity, carbon sequestration, and hydrogeochemical processes); forest landscape change (succession and forest aging); disturbance (insect epidemics, windthrow, and forest fire); habitat provision (habitat suitability and capability, fragmentation, and population dynamics of wildlife); and forest management and planning strategies. Figure 1.2 shows the composition of these topics, in terms of number of studies published.

In addition to the diversification of topics, published knowledge in forest landscape ecology has begun to address related areas. Researchers in this field are advancing ecological concepts, discovering new spatial mapping and analytical techniques, formulating simulation models to extend research hypotheses, and projecting scenarios of spatial processes and patterns. They have begun to explore avenues

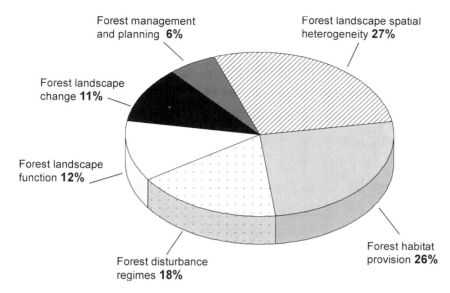

Figure 1.2. Published knowledge in forest landscape ecology has diversified as evident in the composition of the topics in research papers on forested landscapes (total = 170) published in the journal *Landscape Ecology* (1987 to 2005).

for improving forest land-use policies and planning, and to develop applications in support of forest management decisions. The trend of expanded research capacity, specialized subject matter, and increased generation of information is leading to a significant wealth of accumulated knowledge on the ecology of forest landscapes. Even as research knowledge grows, the potential for applications increases, creating the opportunity—and posing the challenge—for knowledge developers to engage in knowledge transfer.

1.2.2. The Potential for Applications

Since the 1980s, a gradual but conspicuous broadening of the goals has occurred in North American forest management driven by various social, ecological, and economic factors (Crow 2002). It is primarily a shift in focus from the supply of timber to the more complex goal of regional sustainability of natural resources, resulting in an associated expansion in forest management planning units from traditional forest stands to larger geographical extents such as ecoregions (Perera and Euler 2000). Attempts to manage forests over broader spatial scales and longer temporal horizons have made forest resource policymakers and managers increasingly aware that a landscape ecological view is necessary to manage toward the goal of forest sustainability.

There are many early examples of this paradigm shift toward a landscape ecological approach in North American forestry. These range from broad legislation in Canada (e.g., the 1994 Ontario Crown Forest Sustainability Act) to environmental assessment processes (e.g., Ontario's 1994 environmental assessment of timber management) to regional plans in the U.S. Pacific Northwest (e.g., Swanson et al. 1990), and to managing for specific conservation values at regional and landscape scales (e.g., spotted owl, Verner et al. 1992, USDA and USDI 1994; old-growth forest, Harris 1984). In addition, forest management planning processes such as the landscape coordination groups commissioned by the Forest Resources Council in Minnesota (Minnesota Statutes 2002) and the Southern Forest Resource Assessment (Wear and Greis 2002) have evolved to rely on landscape ecological approaches. Some jurisdictions, including the province of Ontario, Canada, have explicitly embedded landscape ecological concepts in their forest management directions at all hierarchical planning levels (Table 1.1). Adoption of a landscape ecological view and integration of landscape ecological applications involve substantial growth in the demand for knowledge related to landscape ecology. The question, then, is how the knowledge is incorporated at these levels. To understand this, we need to consider who is (or could be) using the accumulating knowledge base.

1.2.3. Users of the Knowledge

Forest resource managers who develop and operationalize plans to harvest, regenerate, and conserve forest landscapes are the most recognized group of users of forest

Table 1.1. Levels of forest management directions in Ontario, Canada, as an example of a hierarchy that embeds concepts and applications of landscape ecology

Level of forest management directions	Specific articulation	Direction provided	Embedded landscape ecological concepts
Legislation	Crown Forest Sustainability Act (Statutes of Ontario 1995)	Emulating natural forest disturbances as a basis for forest management	A coarse-filter approach to conserving biological diversity
Policy	Old-growth policy for Ontario's Crown forests (OMNR 2003)	Identifying and conserving old-growth forest conditions	Ecoregional heterogeneity in natural disturbances and landscape aging
Guide	Forest management guide for natural disturbance pattern emulation (OMNR 2002)	Using spatiotemporal fire disturbance patterns as a guide to designing harvest patterns	Spatiotemporal variability in crown fire regimes in boreal and near-boreal forest landscapes
Management planning	Forest management planning manual (OMNR 2004)	Managing forests in the context of landscape heterogeneity and dynamics	Long-term forest cover trajectories, wildlife habitat supply, landscape edge, corridors and patch interior

landscape ecology knowledge. However, the realm of potential users of forest landscape ecology knowledge is broad and complex and includes legislators, policymakers, land-use planners, and forest resource managers. Moreover, the decisions of these users are interrelated. They influence the patterns and processes in forest landscapes at various spatiotemporal scales in a nested hierarchy, as illustrated in Figure 1.3. These hierarchical levels are also scale-specific, with different knowledge requirements for their decisions related to forest management. Though most evident in the public sector, these hierarchies of decisionmakers also exist in private sector forest companies. These users may not only have different specific goals in the forest landscape management process, but may also represent differences in a multitude of other traits such as educational backgrounds, institutional cultures, and technological infrastructure, all of which are important in determining whether and how they may use landscape ecology knowledge (Turner et al. 2002).

In addition to those knowledge users who influence forest management decisions directly, many others, loosely referred to as "stakeholders," have an indirect, yet considerable, influence on such decisions. These include recreationists, conservationists, commercial tourist outfitters, public citizen organizations, and environmental nongovernmental organizations, operating at national, regional, or local levels. Such stakeholders are becoming important participants in forest landscape planning processes and, therefore, constitute another group of knowledge users.

Although the exact composition and characteristics of the users of forest landscape ecology knowledge may vary from case to case, all above-mentioned groups collectively play a role in shaping future forest landscapes, and thus represent direct beneficiaries of advances in landscape ecology knowledge. Knowledge developers who are interested in influencing whether and how their knowledge is received and applied will benefit from understanding the roles, goals, and existing knowledge base of these users.

1.2.4. Technological Infrastructure

Another consideration is whether users can accommodate the knowledge base within their technological infrastructure capacity (that is, the technological resources available to the user) and, thus, whether we are at a point at which applications of landscape ecology knowledge are feasible outside the research realm. The past two decades have seen tremendous technological progress in large-scale data-capture methods such as satellite and airborne image recording. As landscape ecology researchers are aware, the accuracy and efficiency of data capture have improved, but data costs have also decreased, making data sources such as Landsat, AVHRR, IKONOS, SPOT, and LiDAR images readily available. Parallel advances in image analysis and GIS software, as well as their increased user-friendliness, coupled with improvements in data storage and computing hardware, have made the use of large-scale information, once accessible only to researchers, increasingly practical and affordable for forest landscape managers. Forest managers in both the public and private sectors are increasingly gaining access to extensive spatial databases of forest

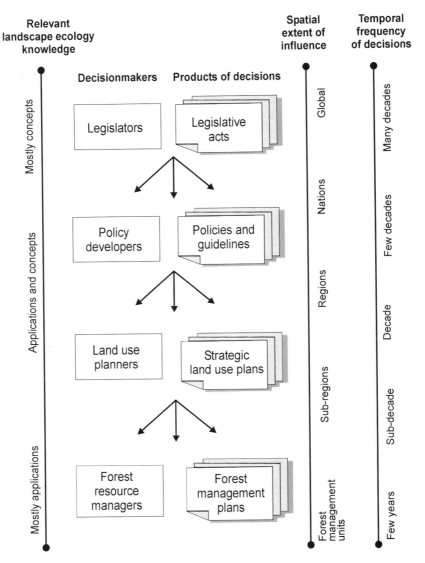

Figure 1.3. An example of the hierarchy of decisionmakers, the level of influence of their decisions in forest management in terms of spatial extent and temporal frequency, and the relevant landscape ecology knowledge.

cover and ancillary information. These databases, regardless of their stage of development, enable forest resource managers to adopt a landscape-level view in their practice, in both a quantitative and a spatially explicit manner. Added to developments in computing and data-acquisition technology is growth in the number of professionals versed in spatial data analysis and computing technology: forest resource management organizations in the public and private sectors are increasingly employing technologists who are adept in using GIS and remotely sensed data in the context of forest management. Therefore, the impact of unavailability and unfamiliarity of spatial data technology, which were considered serious obstacles to applications of forest landscape ecology in the recent past (Perera and Euler 2000; Turner et al. 2002), appears to be diminishing with time.

1.2.5. Barriers to Knowledge Transfer

The factors discussed above can be viewed as a hypothetical source–sink relationship. Accumulating research knowledge is considered to be the source, and the potential application of knowledge is equated to the sink, to which knowledge will be transferred, and the technological and other infrastructure represents the corridor or enabling structure that establishes a link between source and sink and permits the transfer (Figure 1.4). The flow of knowledge depends on the differential between the source and sink and the conductivity of the corridor. Another analog is a supply–demand relationship; that is, demand generated by user applications, the supply of knowledge from research, and a flow of knowledge enabled by the infrastructure that links the two. Both of these analogs of knowledge transfer imply a passive process: because demand is growing, supply is expanding, and the enabling structure is in place, the knowledge is assumed to flow automatically from researchers to practitioners.

The success of such a passive process is predicated on several assumptions about the community of knowledge users. For example, once research knowledge is published, users are assumed to (a) know that knowledge exists, (b) recognize the

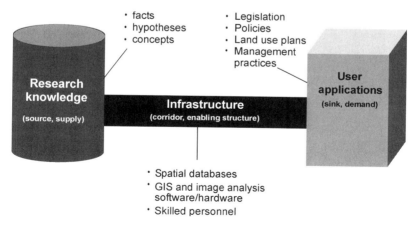

Figure 1.4. An illustration of factors essential for the transfer of forest landscape ecology knowledge and their interlinkages. Some may view knowledge transfer as an automated process analogous to a source–sink or supply–demand relationship. In reality it requires active involvement of knowledge developers and users.

value and relevance of the knowledge, (c) discern the applicability of the knowledge, (d) do the necessary transformation to make knowledge applicable, and (e) if feasible, apply the knowledge appropriately. For these assumptions to be correct, the knowledge developers and the community of users must have a similar philosophical and cultural outlook, similar strategic and tactical goals, and similar scientific and technological environments.

In reality, such similarities rarely exist, invalidating the assumptions about an automated flow of knowledge from developers to users. For example, Turner et al. (2002) identified several generic dissimilarities between the developers of landscape ecology knowledge and managers of natural resources, including incongruity in goals and scales, differences in the nature of the knowledge and data, differences in the training and professional experience of personnel, and differences in institutional culture (Table 1.2).

Furthermore, the forest landscape ecology knowledge generated by developers is innately different from the traditional forest ecology knowledge familiar to forest landscape managers. The resulting differences, some of which are detailed below, may further impede knowledge transfer.

- *Breadth of spatial scale*: Looking beyond the level of forest stands to address forest regions, which are the basis of forest landscape ecology, is not natural to forest managers and planners, and it may not even be accommodated by the present policy and socioeconomic frameworks.
- *Multidisciplinary complexity*: The breadth of the spatial scale results in inevitable social and economic ramifications at the outset of any application of landscape ecology knowledge, and this necessitates broader considerations, often across multiple research disciplines, than has been customary.

Table 1.2. Major differences between landscape ecology researchers and forest managers (adapted from Turner et al. 2002) that prevent an automated knowledge flow from developers to users

	Landscape ecology researchers	Forest managers
Goals	Understand causes and ecological consequences of spatial heterogeneity	Maintain or alter natural resources for societal objectives as guided by local, state, and federal statutes
Scales	Ecologically meaningful scales	Management-oriented scales
Tools/methods	Spatial modeling and analysis, geographic information systems, experiments	Harvest, prescribed fires, wildlife management, restoration, habitat manipulation
Training/experience of personnel	Training in ecology, no management experience	Outdated or little training in ecology, rich management experience
Data	Observation results, simulation results, experimental results, remote sensing data	Observation results, remote sensing data
Institutional culture	Publish or perish	Crisis control and problem-solving

- *Length of temporal scale*: A single forest harvest rotation is the most common planning horizon for forest managers, dictated by economic realities, but forest landscape ecology addresses longer-term planning horizons.
- *Stochasticity in broad landscape processes*: Using traditional knowledge, forest managers often consider determinism to be the de facto status at broad scales, which supports only one trajectory of structure and composition in designing future landscapes, whereas forest landscape ecology may introduce alternative outcomes.
- *Reliance on conceptual models*: Use of predictive and prescriptive models is the norm in forest management, and this makes the more abstract scenario-simulation models and exploratory models that are designed to provide insight and context in landscape ecology unfamiliar to these potential users of the technology.
- *Focus on coarser resolution*: Stand-level and finer resolution of information is the staple input to forest management planning, whereas forest landscape ecology relies on resolutions coarser than this level; as a result, forestry practitioners may question the value of this information.

In addition, forest landscape ecology knowledge may not be readily available to and usable by practitioners. This problem may arise from causes such as the following.

- *Lack of awareness*: Forest landscape managers may not be aware of the accumulating knowledge base, which may be mostly available in journals meant for researchers, and may not understand its relevance to their management practices.
- *Usability of knowledge*: Much of the forest landscape ecology knowledge is still available only in its primary form, such as in complicated models that rely heavily on complex computing technology, rather than in the form of user-friendly tools and applications. This makes direct use of the knowledge difficult for forest managers.
- *Incompatibility with their needs*: Even where applicable knowledge suitable for its intended user is available, it may have been developed without considering the user's specific goals.
- *Incompatibility with existing infrastructure*: Even when applicable knowledge is compatible with the user's needs and goals, users may find that the applications are not compatible with their present suite of applications, databases, and computing technology.

Given these impediments, it is obvious that passive knowledge transfer will not occur in most instances, and that reliance on a solely automatic process of knowledge transfer is likely to widen the disparity between the volume of generated knowledge and the successful application of this knowledge in forest landscape management.

1.3. WHAT CAN FOREST LANDSCAPE ECOLOGISTS DO TO ADVANCE KNOWLEDGE TRANSFER?

All developers of forest landscape ecology knowledge, whether they are academics, researchers, or technologists, have the capability to actively engage in knowledge transfer, albeit to varying degrees. This involvement in knowledge transfer would help ensure that the knowledge they work to generate has an opportunity to be applied appropriately. The broad goal of transfer is to make users aware of the knowledge available and its appropriate application, as well as to impart the technological skills required to apply that knowledge.

1.3.1. Understand the Basics of Knowledge Transfer

Knowledge transfer is an essential step to ensure timely and effective application of knowledge already developed, as well as to identify future needs. The nature of the knowledge to be transferred ranges widely and varies with the circumstances. At one extreme, the principles and concepts of landscape ecology are required for users to understand underlying forest landscape patterns and processes, which provide the context for the necessity and appropriateness of landscape ecological applications. At the other extreme, many skills and technological knowledge are required to understand and use the models, user tools, and spatial data.

Knowledge developers (academics, researchers, or technologists) may transfer knowledge through direct contact with users (legislators, policy developers, land-use planners, forest resource managers, or stakeholders). The specific goal of a transfer activity may range from creating awareness of an emerging concept among users, educating users about the meaning and potential use of a specific research finding, or training users to use a new tool; more than one of these goals may be achieved simultaneously. The intended outcome also ranges widely, from knowing about a concept to understanding the principles and interrelationships with other factors and appropriately applying the new concept or tool. An important aspect of this engagement is that it is reciprocal: users provide feedback to developers about the transfer they received, or initiate transfer and future research by articulating their needs to the knowledge developers. In some instances, professionals trained specifically in knowledge transfer may enter the process and participate in the transfer; these individuals are often referred to as "extension specialists." For the purpose of this chapter, we use the generic term *transfer specialist* for these professionals. Working definitions for relevant transfer-related terms, and commonly used synonyms, are provided in Table 1.3.

As we noted earlier, there are subtle differences between the terms *knowledge transfer*, *technology transfer*, and *extension*. Here, we have used the term *knowledge transfer* to mean the broad group of activities that will increase the understanding of landscape ecology principles, concepts, and specific facts by users, through education, thereby providing a basis for applying the knowledge. Knowledge transfer is

Table 1.3. Working definitions and examples for commonly used knowledge transfer terms

Term	Working definitions	Related terms and examples
Who is involved?		
User	An individual or a group that interacts with developers or transfer specialists to (a) receive knowledge for application and (b) provide feedback on their needs and the applicability of the knowledge	Audience, user, client, stakeholder, forest manager, policymaker, legislator, land-use planner
Developer	An individual or a group that (a) generates knowledge or technology for application by practitioners and (b) receives feedback from practitioners	Academic, researcher, technologist
Transfer specialist	An individual or a group that interacts with developers and practitioners to enhance and expedite the knowledge transfer process	Extension specialist, transfer professional, research liaison, GIS specialist, GIS technologist
What is being transferred?		
Knowledge	Generalized principles, concepts, and specific facts that provide the contextual basis for application	Research findings, models, decision-support systems, methods
Technology	Mechanical means necessary for the application of knowledge	Techniques, user tools, information, data
How accomplished?		
Engagement	Direct interaction among developers, practitioners, and transfer specialists to enable awareness, education, training, and feedback	Involvement, cooperation, collaboration
Awareness	Developers and transfer specialists increasing the practitioner's cognizance of knowledge and technology	Transfer, extension, outreach
Education	Developers and transfer specialists imparting knowledge through activity planned to increase understanding	Transfer, extension, outreach
Training	Developers and transfer specialists helping practitioners to learn a technology or skill through instruction and guided practice	Hands-on exercises, guided practice
Feedback	Developers and transfer specialists receiving practitioner response to transfer activities and becoming more aware of practitioner needs	Evaluation

also a precursor to technology transfer, which encompasses the broad group of activities that increases the users' awareness of applications of knowledge and their skills in using specific tools through training. The chronological progression is from awareness to understanding and finally to applying the knowledge. The overall goal of the suite of transfer activities, commonly referred to as extension, is to help users progress toward the goal of successfully applying landscape ecology knowledge. These principles provide a general overview of the basic elements of knowledge transfer and their interlinkages, and can be applied to both individuals and organizations. The mechanisms of how the various components interact are not addressed here. For more details on these interactions, interested readers are directed to, for example, Argyris and Schön (1978), Reed and Simon-Brown (2006), and Rogers (1995).

In Figure 1.5, we illustrate a hypothetical scenario in which the principles discussed above are put into practice to transfer a landscape ecology model to practitioners. In this example, a spatial research model is converted into an application, an exercise that requires adapting the model to a user-friendly GIS-based tool, while capturing local knowledge. The knowledge transfer process in this instance is com-

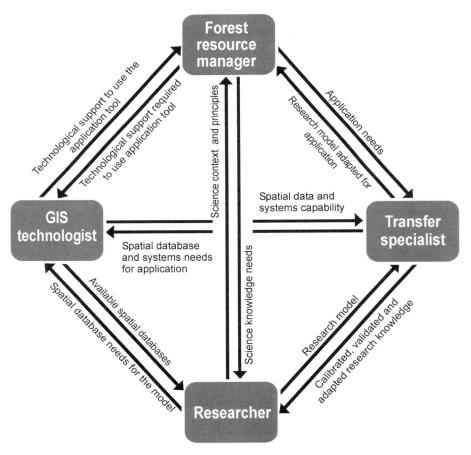

Figure 1.5. A hypothetical scenario in which a landscape ecological model is developed by researchers and converted into a locally adapted application with support from transfer specialists and GIS technologists to meet the needs of forest resource managers. Ideally, all participants engage and interact as a group, rather than in isolated pairs.

plex, and requires both local ecological expertise and GIS technological expertise that goes beyond the developer's understanding of user needs and the user's understanding of scientific principles. The participants are engaged in two-way communication and each plays a vital role in the process. (In practice, the engagements and communications may not occur in pairs: ideally, all participants engage and interact simultaneously as a group). We do not imply that transfer specialists and GIS specialists are absolutely necessary in all cases to transfer knowledge from developers to users; the developers may perform the additional role of transfer specialists and the users the role of GIS specialists.

1.3.2. Play an Active Role

As we noted above, many participants may take part in the process of transferring landscape ecology knowledge. However, a major share of the responsibility for making knowledge transfer an active process rests largely with the developers of knowledge, and they can take on this role by promoting the flow of knowledge between themselves and the users. We recognize three knowledge transfer approaches that account for differences in the role of the knowledge developers. The hypothetical model we presented earlier (Figure 1.4) can be modified to fit these approaches.

First is the supply-driven ("push") transfer approach, in which the developer initiates and powers the knowledge flow. For example, in relation to Figure 1.4, knowledge developers prime the flow of knowledge and drive the knowledge transfer process by proactively creating awareness and educating users. This is analogous to marketing of knowledge. The role landscape ecologists play in this approach must not be confused with environmental advocacy: rather than *advocating* research results and outcomes, the developer *creates awareness* of principles and opportunities. The goal is to make users aware of new scientific concepts, research findings, approaches, methods, and techniques by means that extend beyond publishing in peer-reviewed journals or presenting papers at scientific meetings. We contend that this approach is particularly necessary and effective for broader-level users in the hierarchy in Figure 1.3, who deal with longer-term issues at global, national, and regional scales. There are many examples of this approach in various aspects of ecology, in which scientists have successfully created awareness among legislators, policymakers, and resource managers. Examples such as the emergence of forest landscape management philosophies, including biodiversity conservation, emulating natural disturbance, conserving old-growth forests, and the adoption of practices such as the provision of wildlife corridors and forest patch interior can be attributed to supply-driven transfer.

Second is the demand-driven ("pull") transfer approach, in which users initiate knowledge flow. In relation to Figure 1.4, they prime the knowledge transfer process by recognizing the need for scientific answers to resource management problems. However, the participation of knowledge developers in this approach is no less important than their role in the supply-driven approach. It is analogous to suppliers responding to high demand in a specific market: Knowledge developers must recognize and respond to the specific needs of users, and must provide the necessary science-based solutions, sometimes by means of focused research to solve specific problems identified by the users. Our view is that demand-driven knowledge transfer is effective with users at finer levels of the hierarchy in Figure 1.3—those who plan and manage forest landscapes—mostly in the context of legislation, policies, and other established broader-level directions. Most examples of this transfer approach are models, tools, and decision-support systems developed to meet forest landscape planning or management needs. The popular use of strategic forest landscape planning and harvest-design tools, models of forest succession and disturbance dynamics, tools for assessing landscape patterns, and models of habitat supply and population dynamics can be attributed to demand-driven transfer.

The third approach is more balanced, in which the knowledge flow occurs as a result of both the push from developers and the pull from users—that is, as a result of both the supply of knowledge and demand for its use. This represents a collaborative and iterative approach (Fall et al. 2001, Ruhleder and Twidale 2000), in which the role of landscape ecologists is a continuous engagement in transfer and in developing successive iterations of a product to incorporate the experience and perspectives of everyone involved. The collaborative-iterative approach, in its ideal formation, does not fit the model in Figure 1.4 because there is little separation between developers and users, and no distinct push or pull to prime the process. Principles of adaptive learning, though discussed in the context of natural resource management institutions by Stankey et al. (2005), also may apply here, where knowledge exchange and learning are iterative. Presently, instances of this transfer approach are relatively rare because it (a) requires users to be relatively well versed in landscape ecology knowledge and familiar with the developers and (b) requires developers to be familiar with users and forest landscape management. However, with time, as the transfer process for forest landscape ecology knowledge matures, the collaborative-iterative transfer approach is likely to become increasingly popular.

If landscape ecological knowledge transfer is viewed in terms of an evolutionary process, then the supply-driven transfer approach can be viewed as the most primitive, where knowledge developers initiate the process by creating awareness of landscape ecology knowledge and its potential for applications among users. As landscape ecology knowledge transfer evolves, the user's increasing awareness of the knowledge creates a pull (a demand) for potential applications, leading to a stage in which transfer is initiated by user demand. Once started by either push or pull, the momentum of the transfer process could be driven and maintained by a combination of user demand and knowledge supply. In the final stage of evolution, when knowledge developers and users are mutually familiar with the knowledge and knowledge development capacity and with the applications and user needs, the transfer moves to a collaborative-iterative mode.

1.4. SUMMARY

Forest landscape ecology is maturing as a discipline and its knowledge base is rapidly expanding. A necessary next step is to ensure appropriate application of this accumulating knowledge in forest management. For this to occur, it is imperative that landscape ecologists become familiar with knowledge transfer, which is a process in which developers interact with users, and make them aware of the available knowledge and its appropriate use. As well, developers learn user needs, which promote discovery and iterative improvement of applications.

Knowledge transfer can occur in many different ways; no one standard method is universally suitable. Approaches effective for introducing new landscape ecological concepts may not work as well for encouraging the adoption of new technology. Regardless of the approach, it is evident that developers have an active and leading role in the process. Fulfilling this role requires an understanding of basic knowledge transfer concepts, principles, and practices and the willingness to engage with users.

Although successful transfer is an essential prerequisite, it alone cannot ensure that landscape ecological concepts are appropriately applied for a myriad of practical reasons. Still, the opportunity awaiting researchers to convey their findings to users for application is vast and timely. By considering knowledge transfer as an integral part of their activities, forest landscape ecologists have the opportunity to move their field of study from the abstract to an applied discipline.

LITERATURE CITED

Argyris C, Schön D (1978) Organization learning: a theory of action perspective. Addison–Wesley, Reading.
Crow TR (2002) Putting multiple use and sustained yield into a landscape context. In: Liu J, Taylor WW (eds) Integrating landscape ecology into natural resource management. Cambridge University Press, Cambridge, UK pp 349–365.
Fall A, Daust D, Morgan D (2001) A framework and software tool to support collaborative landscape analysis: fitting square pegs into square holes. Trans GIS 5(1):67–86.
Franklin JF, Forman RTT (1987) Creating landscape patterns by forest cutting: ecological consequences and principles. Landscape Ecol 1:5–18.
Gutzwiller KJ (ed) (2002) Applying landscape ecology in biological conservation. Springer, New York.
Harris LD (1984) The fragmented forest: island biogeography theory and the preservation of biotic diversity. University of Chicago Press, Chicago.
Liu J, Taylor WW (eds) (2002) Integrating landscape ecology into natural resource management. Cambridge University Press, Cambridge, UK.
Minnesota Statutes (2002) Minnesota Sustainable Forest Resources Act. Statute MS 89A. Gov't of Minnesota, St. Paul.
Mladenoff DJ, Baker WL (1999) Advances in spatial modeling of forest landscape change: approaches and applications. Cambridge University Press, Cambridge, UK.
OMNR (2002) Forest management guide for natural disturbance pattern emulation. Version 3.1. Ont Min Nat Resour, Toronto.
OMNR (2003) Old growth policy for Ontario's Crown forests, V. 1. Ont Min Nat Resour, For Pol Ser, Sault Ste. Marie.

OMNR (2004) Forest management planning manual for Ontario's Crown forests. Ont Min Nat Resour, Sault Ste. Marie.
Perera AH, Euler DL (2000) Landscape ecology in forest management: an introduction. In: Perera AH, Euler DL, Thompson ID (eds) Ecology of a managed terrestrial landscape: patterns and processes of forest landscapes in Ontario. UBC Press, Vancouver, pp 3–12.
Perera AH, Euler DL, Thompson ID (eds) (2000) Ecology of a managed terrestrial landscape: patterns and processes of forest landscapes in Ontario. UBC Press, Vancouver.
Perera AH, Buse LJ, Weber M (eds) (2004) Emulating forest landscape disturbances: concepts and applications. Columbia University Press, New York.
Reed AS, Simon-Brown V (2006) Fundamentals of knowledge transfer and extension. In: Perera AH, Buse LJ, Crow TR (eds) Forest landscape ecology: transferring knowledge to practice. Springer, New York.
Rochelle JA, Lehmann LA, Wisniewski J (eds) (1999) Forest fragmentation: wildlife and management implications. Brill Press, Leiden, The Netherlands.
Rogers EM (1995) Diffusion of innovations. 4th ed. The Free Press, New York.
Ruhleder K, Twidale M (2000) Reflective collaborative learning on the Web: drawing on the master class. First Monday 5(5). (http://www.firstmonday.org/issues/issue5_5/ruhleder/)
Stankey GH, Clark RN, Bormann BT (2005) Adaptive management of natural resources: theory, concepts, and management institutions. USDA For Serv, Pacific Northwest Res Stn, Portland. Gen Tech Rep PNW-GTR-654.
Statutes of Ontario (1995). Crown Forest Sustainability Act, revised. R.S.O. 1998. Chapter 25 and Ontario Regulation 167/95.
Swanson FJ, Franklin JF, Sedell JR (1990) Landscape patterns, disturbance, and management in the Pacific Northwest, USA. In: Zonneveld IS, Forman RTT (eds) Changing landscapes: an ecological perspective. Springer-Verlag, New York, pp 191–213.
Turner MG, Crow TR, Liu J, Rabe D, Rabeni CF, Soranno PA, Taylor WW, Vogt KA, Wiens JA (2002) Bridging the gap between landscape ecology and natural resource management. In: Liu J, Taylor WW (eds) Integrating landscape ecology into natural resource management. Cambridge University Press, Cambridge, UK, pp 433–465.
USDA Forest Service and USDI Bureau of Land Management (1994). Record of decision for amendments for Forest Service and Bureau of Land Management planning documents within the range of the northern spotted owl. USDA Forest Service, Pacific Northwest Region, Portland, OR, USA.
Verner J, McKelvey KS, Noon BR, Gutiérrez RJ, Gould GI Jr, Beck TW (tech coord) (1992) The California spotted owl: a technical assessment of its current status. USDA For Serv, Pacific Southwest Res Stn, Albany, CA. Gen Tech Rep PSW-GTR-133.
Wear DN, Greis JG (2002) The southern forest resource assessment: summary of findings. J For 100(7):6–14.

2

Transfer and Extension of Forest Landscape Ecology: A Matter of Models and Scale

Anthony W. King and Ajith H. Perera

2.1. Introduction
2.2. What Are Forest Landscape Ecological Models?
2.3. Who Uses Forest Landscape Ecological Models?
2.4. What Makes Forest Landscape Ecological Simulation Models Less Appealing to Users?
    2.4.1. Unfamiliarity with Topic
    2.4.2. Uncertainty about the Purpose of Simulation Models
    2.4.3. Unclear Assumptions and Application Limitations
    2.4.4. Dissatisfaction with Abstractions and Assumptions
    2.4.5. Discomfort with Modeling at Large Scales
    2.4.6. Discomfort with Stochasticity and Variability in Simulated Processes
    2.4.7. Distrust of "Black Box" Models
    2.4.8. Distrust of Methods of Model Validation
    2.4.9. Unavailability of Computing Technology and Spatial Data
    2.4.10. Necessity for Third Party Involvement
2.5. Is Misunderstanding of Scale a Serious Impediment to Users of Landscape Ecology?
    2.5.1. Have We Discussed Enough about Scale?
    2.5.2. Has the Discussion of Scale Been Clear?
    2.5.3. Has the Theory of Scale in Ecology Been too Esoteric?
    2.5.4. Has There Been too Much Focus on Multiple Scales?
    2.5.5. Has There Been too Little Attention to the Defining Principles of Landscape Ecology?
    2.5.6. Has There Been Sufficient Synthesis?
    2.5.7. Perhaps It's too Early?

ANTHONY W. KING • Oak Ridge National Laboratory, Environmental Sciences Division, Bldg 1509, MS 6335, Oak Ridge, TN 37831, USA.   AJITH H. PERERA • Ontario Ministry of Natural Resources, Ontario Forest Research Institute, 1235 Queen St. E., Sault Ste. Marie, ON P6A 2E5, Canada.

2.6. How Can Researchers Make Simulation Models More Appealing to Users?
   2.6.1. Understand the User, Not Only the Use of the Models
   2.6.2. Develop Simple Models
   2.6.3. Clarify the Limitations of the Model to Its Users
   2.6.4. Transfer Knowledge to Users Interactively, before, during, and after Model Development
   2.6.5. Synthesize an Understanding of Scale from the Perspective of Model Application
2.7. Mutual Benefits
   Literature Cited

2.1. INTRODUCTION

Forest landscape ecology involves examining relationships in spatial geometry among forest elements at broad spatial and temporal scales and higher levels of ecological organization—whether the focus is on physical processes such as hydrology, biological functions such as primary productivity, biophysical processes such as forest fire, or human activities such as forest harvest. In short, forest landscape ecology is a subset of the more general field of landscape ecology, which seeks to understand how spatial patterns and relationships influence forest process. Although, in principle, an understanding of how spatial relationships among individual trees in a stand influence stand growth and productivity could qualify as forest landscape ecology, in practice, the spatial extents of forest landscape ecology are much larger than forest stands; they involve large watersheds and geographical regions. Hence, forest landscape ecology, as used here, should be understood as the study of how spatial patterns and interactions influence the processes and dynamics of heterogeneous forested areas much larger than homogeneous stands of even-aged trees. The science of landscape ecology is defined primarily by its focus on how spatial patterns and interactions influence ecological process, not solely by spatial extents that are large from a human perspective. (At least, that is how it *should be* defined; the point is debated within the community of landscape ecologists.) Nevertheless, when applied to forests, forest landscape ecology deals almost exclusively with large spatial extents.

These large spatial scales are also generally associated with longer time scales, and the temporal domain of forest landscape ecology is much longer than the lifespan of individual trees or even individual stands; it extends toward the time scale of changes in the biogeographical distribution of forests. Moreover, the spatial and temporal resolutions are much coarser than those typically considered in traditional forest ecology. Consequently, the knowledge and information developed in forest landscape ecology addresses broader and coarser spatial and temporal scales than are familiar to the policymakers, planners, and practitioners involved in forest management at national, regional, or local levels. Ironically, it can be argued that the questions and information needs of these "end users" require a consideration of these larger scales, but in our experience, end users of forest landscape ecological knowledge are more accustomed to the smaller scales of traditional forest ecology. This results in a mismatch between the scales of the problems, the scales addressed by the science, and the scales understood by the users. It is not surprising that transfer of knowledge generated by the science of landscape ecology into forestry applications is at best uneven.

Another consequence of the breadth of the spatial and temporal scales in forest landscape ecology is the impracticality of developing scientific knowledge by traditional experimentation. For example, even forested watersheds, which are large relative to the size of most field experiments, may be small relative to the scales of forest landscape ecology. For this reason, simulation models have become an essential research tool for forest landscape ecologists. Such models are evident in all aspects of forest landscape ecological research: studies of forest landscape composition, structure, dynamics, and function, as well as their management. Invariably, these models are also the primary means by which landscape ecological knowledge is conveyed for applications in forest management. This is different from the mostly empirical ecological knowledge traditionally available to forest resource managers.

Given this background, our goals in this chapter are to examine and illustrate the generic barriers to popular use of forest landscape ecological models and to offer potential solutions. We will focus less on what is wrong with landscape ecological models (an advice typically offered to advance modeling concepts), and more on how researchers can make models more attractive to forest resource managers by understanding the user's perspective. Consideration and understanding of the scale of forest landscape ecology are important for both the modelers and the managers. Misunderstanding or miscommunication of the principles of scale could hinder the transfer of knowledge (e.g., models) generated by forest landscape ecology to users. Accordingly, we consider some of the common barriers to understanding the scale of forest landscape ecology that could impede the widespread use of models. Our comments are primarily focused on the communication of scale by researchers and its understanding by managers, but we also identify problems of scale in ecological models and explain the implications of scale, when necessary, to clarify the potential for miscommunication or misunderstanding.

Our presentation is not a comprehensive review of literature—the body of literature on models and scale from the perspective of applied landscape ecology is limited—rather, it is a synthesis of our experiences and insight gleaned from a combined several decades of research in scale, landscape ecology, and forest modeling and our interactions with potential users of landscape ecology in forest management and other applications.

2.2. WHAT ARE FOREST LANDSCAPE ECOLOGICAL MODELS?

Before we discuss the transfer of forest landscape ecological models and their applications to users, let us examine what a *model* means in this context. It is used in landscape ecological parlance with a variety of mathematical, statistical, biological, and social connotations, and the literature is replete with model definitions and descriptions. For example, in a recent discussion of ecological models for resource management, Dale (2003a) offered three broad groupings of landscape ecological models: heuristic (conceptual abstractions showing interrelationships among variables), physical (scaled-down expressions of the real world in two or more dimensions), and mathematical (descriptions of numerical interrelationships among variables). In the context of this chapter, we limit our discussion to simulation models, which are a subset of Dale's mathematical models in which modelers use numerical and computational methods to describe and investigate the behavior of the system being modeled. Simulation models in forest landscape ecology may be developed for various reasons, ranging from exploring (e.g., examining what-if scenarios, in which the known variables or their values and functions can be changed), to predicting (e.g., projecting specific outcomes based on a specific set of known variables and functions). Regardless of these variations, a simulation model, in essence, is a logical and an explicit articulation of an abstracted relationship between known ecological variables and unknown ecological variables. This articulation is quantitative; unknowns are expressed or simulated as a function of known variables and valid only under a given set of circumstances (i.e., the model assumptions). Figure 2.1 describes, in abstract, the essential anatomy of a simulation model in forest landscape ecology.

In principle, a forest landscape ecological model could be any model of the forested landscape at large spatial and temporal extents (as defined above). In

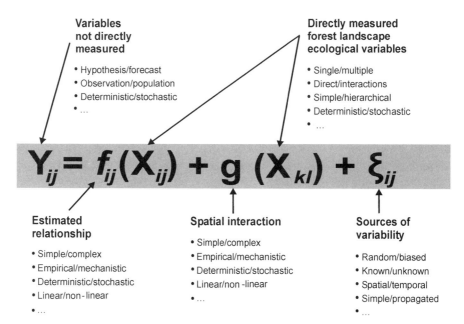

Figure 2.1. Anatomy of the basic components of a forest landscape ecological simulation model. i,j represents a two-dimensional index of spatial variability in model components, and kl represents the influence of spatial location on ij.

practice, forest landscape simulation models tend to focus on variation within those extents and generally disaggregate the landscape into patches, a mosaic of polygons, or a regular geometrical grid of squares, triangles, or hexagons. Thus, most of the considerations that must be addressed in traditional nonspatial or aspatial forest simulation models are compounded by considerations of how measured variables, relationships between known (measured) and unknown (modeled), and sources of variability vary in space, or in the models from cell to cell in the grid or mosaic (Figure 2.1). In addition, forest landscape ecological models must, or should, consider interactions between variables at one location and those at others. Forest landscape models sometimes ignore these interactions, or at least presume they are inconsequential. However, a forest landscape ecological model should at least make this latter assumption explicit if it is to be true to its heritage as a model informed by the science of landscape ecology.

Simulation models differ widely in their variables, assumptions, and functions and these differences may be evident in specific features of model components. Directly measured variables may be single or multiple, deterministic or stochastic, and simple or hierarchical. When applied to a landscape, the spatial variation or the aggregation of spatial variability in these observed variables differs among models. The estimated relationships may be simple or complex, empirically derived or mechanistically constructed, deterministic or stochastic, and linear or nonlinear. The parameters describing the relationships may vary spatially, and each function may change from one location to the next. Sources of variability or "error" in a model can be random or biased, known or unknown, and simple or propagated, and the variability may or may not change through time. When applied to a landscape, spatial variation of these sources may or may not be explicitly considered in the model. More generally, forest landscape ecological models differ considerably in how spatial variation in observed variables, relationships, modeled variables, and sources of error are explicitly represented or spatially aggregated. Forest ecological models can differ greatly; forest *landscape* ecological models differ even more. Regardless of these differences, all forest landscape ecological models can be expressed using the abstraction in Figure 2.1.

Forest landscape ecology offers many different research topics in which model development is common. These range from physical and biological processes to anthropogenic processes, and include models that focus on, for example, hydrology, climate change, forest fires, carbon sequestration, metapopulations, forest succession, harvesting, and urbanization. These different focus areas often lead to variation in how the elements in Figure 2.1 are represented in the models.

2.3. WHO USES FOREST LANDSCAPE ECOLOGICAL MODELS?

Forest landscape ecological models are diverse with respect to their variables and mathematical formulations (Figure 2.1), as well as their use. In addition to their many different scientific uses (e.g., as a heuristic, as a framework for synthesis and integration, to calculate quantities, as spatially explicit hypothesis generators, or to test hypotheses), researchers in forest landscape ecology also use simulation models to integrate and extend ecological information for applied use in forest landscape management (e.g., Buongiorno and Gilless 2003; Jansen et al. 2002; Mladenoff and Baker 1999; Perera et al. 2004). They are intended for use in design, planning, and managing forest landscapes either by generating broad contextual information or by providing answers to what-if questions raised under specific management scenarios. Table 2.1 lists an example set of uses of forest landscape simulation models in management, ranging from legislation to harvest planning.

Simulation models that focus on topics such as climate change, carbon sequestration, metapopulation dynamics, forest fire regimes, urbanization, and pest epidemics provide contextual information to aid in the development of strategic policies and plans for forest management. Primary users of such models are at the higher end of the decisionmaker hierarchy, and range from legislators and policymakers to land-use planners who focus on larger spatial extents and longer time horizons. Simulation models that focus on topics such as forest succession, habitat supply, and harvesting may provide answers to questions raised during forest management planning and decisionmaking at a tactical level. Primary users of such models are forest resource managers who focus on (relatively) smaller spatial extents and shorter time horizons (Table 2.1).

Although these groupings overlap, recognition of the end uses of the model is an *a priori* need for successful model development and transfer of the model to its users. For example, the "push" transfer approach (see Perera et al. 2006) is more effective with legislators and policymakers, because researchers generate scientific knowledge to provide contextual awareness and baseline information in response to forest landscape ecological issues. Some examples of push-based knowledge transfer include the results of metapopulation models, global and regional climate change models, and models of invasive species. On the other hand, the "pull" transfer approach (see Perera et al. 2006 for details) appears to be more common with forest resource managers; in this approach, researchers develop simulation models based on user demand for tools, including decision-support systems. Some examples of pull-based knowledge transfer include harvest simulation models, fire-spread models, and forest succession models. We do not imply that this dichotomy is appropriate in all situations: in most cases a combination of the two transfer approaches may be appropriate. And the push approach, if effective, will often lead to a pull. When push switches to pull, the modeling requirements are likely to change because the models are developed to meet different needs.

2.4. WHAT MAKES FOREST LANDSCAPE ECOLOGICAL SIMULATION MODELS LESS APPEALING TO USERS?

In this section, we examine common barriers to the application of simulation models. Our discussion is embodied in the statement that "we cannot expect people to apply ideas that they do not understand or support" (Gutzwiller 2002a). We found

Table 2.1. Examples of uses of forest landscape ecological simulation models for decisionmaking by a range of users from legislators to forest managers

Decisionmaking hierarchy in forest landscape management	Products of decisions	Spatial extent of influence	Temporal window of decisions	Examples of areas of application	Examples of forest landscape ecological simulation models
Legislators	Legislation	Global, national, and regional	Many decades	Endangered species acts, climate change	Metapopulation models Global and regional circulation models
Policy developers	Policies and guidelines	National and regional	Several decades	Old-growth conservation, emulating disturbance, invasive species	Forest fire regime models Species migration and diffusion models
Forest land-use planners	Forest land-use strategic plans	Regional and subregional	A single decade or subdecades	Design of parks and protected areas, wildlife corridors and habitat supply, watershed management	Watershed models Habitat supply and population models Forest fragmentation and urbanization models
Forest resource managers	Harvest and regeneration tactical plans	Forest management units	A few years	Timber supply, forest fire management, forest pest management	Harvest planning models Fire-spread models Pest epidemic models

that the literature on simulation models and model development does not commonly address this topic: few authors address problems that users face in applying landscape ecological models (e.g., Dale 2003a,b; Gutzwiller 2002b; Perera and Euler 2000; Turner et al. 2002). Their views are summarized in the ensuing discussion, augmented by a synthesis of our own insight gained during decades of experience in model development and transfer to forest resource managers and policymakers.

2.4.1. Unfamiliarity with Topic

The broad spatial and temporal scales of the concepts addressed in forest landscape ecological simulation models are new and exotic to most potential users of these models because these topics were not part of their training. Though this obstacle is temporary, and will gradually disappear with the turnover among forestry professionals, it has been a significant impediment to ready transfer of modeling knowledge to forest resource managers and policymakers and will continue to be for some time. As such, much of the early effort in transferring models involves increasing user familiarity with scale and spatial concepts. These issues are discussed in more detail in Section 2.5.

2.4.2. Uncertainty about the Purpose of Simulation Models

Users, especially those with tactical end-use goals, may expect simulation models to generate information that they can use directly in decisionmaking. Though this is a reasonable expectation, not all simulation models are intended to be decision-support systems. If they are not designed to assist or support specific decisionmaking under predetermined circumstances, attempts to use simulation models to support management decisions can be futile or even counterproductive. Similarly, exploratory models that attempt to reveal emergent properties and provide contextual information must not be used to predict or forecast scenarios for decisionmaking. Such occurrences, not uncommon in our experience, are mostly a result of ambiguity in the model developer's elucidation of the model's purpose.

It is not surprising that users misunderstand the purpose of a simulation model or of modeling in general. Modelers themselves often misunderstand or remain unaware of the diverse purposes behind model development and use and the implications of this diversity. In that context, communication of purpose can obviously be difficult and limited. Modelers have their own biases that blind them to the differences in model specifications required by differences in the model's purpose. Uncertainty and differences of opinion among researchers about the definition of decision-support systems, about how models can support decisions, and about who will use the models to make decisions, contribute to the general confusion about the purpose of simulation models.

2.4.3. Unclear Assumptions and Application Limitations

Even if the model purpose is made clear to users, the specific assumptions, scales, and other premises of simulation models may remain unclear. This occurs frequently when model developers fail to unambiguously explain the model's assumptions and limitations to its users. Developers generally understand the assumptions and limitations of their models, but often fail to make those assumptions and limitations explicit. It is worth noting that researchers who use models developed by others and who may then promote the use of those models through transfer to an application (e.g., management) often do not understand all the assumptions and limitations of the model themselves. The chain of communication between model developers, scientific users of the model, and managers or decisionmakers who apply those merits careful attention.

Though there is no assurance that clear communication of assumptions and limitations will lead to correct use of models, or that correct use will lead to correct policies and management decisions, unclear communication can easily lead to misuse of models. This, in turn, will certainly lead to incorrect or inappropriate—or at least ill-informed—policies and management decisions. As a result, users will lose confidence in simulation models, and this lack of confidence will become a serious long-term impediment to the application of these and other models in forest management.

2.4.4. Dissatisfaction with Abstractions and Assumptions

Often, users of models are uncomfortable defining forest landscape ecological systems based on explicit assumptions, which they believe artificially reduce real-world complexity. In fact, many model developers are also uncomfortable with simplified models. But simulation models are designed to be abstracted representations of ecological systems, with the abstraction governed by strict assumptions, and do not, should not, and cannot address all details of the systems they model. Attempts by some model developers to produce parsimonious models, which emphasize an economy of explanation in conformity with Occam's razor, pose a problem for users who expect models to address all possible details of forest landscape structure and function. At the policy design level, attempting to address social, biological, and economic aspects through modeling is a daunting task for modelers, and for users, even when modelers succeed in developing such complex models. But even an agreement that a model should be parsimonious while still meeting the stated objectives does not guarantee that dissatisfaction with the model's abstractions and assumptions will be avoided. Different understandings of and biases about what is important, what is essential, and what is "just" detail arise from different scientific and management perspectives and can lead to different abstractions, albeit parsimonious, which may not be readily acceptable.

2.4.5. Discomfort with Modeling at Large Scales

Forest landscape simulation models address concepts, use data, and produce simulated scenarios at scales that lie beyond the typical scale of human perception, and this makes understanding of the models difficult. The use of maps and remote sensing helps, but simulation models of large-scale patterns (e.g., species extinction) that use "coarse" data such as satellite imagery or simulation models of low-probability events of occurrence such as infrequent incidence in time (e.g., flooding) or rarity in space (e.g., forest fires), force users to address unfamiliar spatial and temporal extents and intervals and equally unfamiliar resolutions. Similarly, models of large-scale spatial processes that play out only over a long time scale (e.g., climate change and biogeographical redistribution of species) force users to deal with unfamiliar time periods that may well exceed the accustomed scales of forest management. Although we have found that users can implicitly use and synthesize broad-scale information, doing so explicitly and quantitatively through simulation models appears more complicated. This is somewhat ironic for those who work with forests, since managers understand that trees often live relatively long, and the forests they occupy can persist relatively unchanged for multiple human lifetimes. Forest ecologists and managers are accustomed to "standing among the trees to see the forest," but the different scales and perspectives of landscape ecology and forest management make the larger scales and perspectives unfamiliar to individuals with more traditional experience and training. This lack of familiarity generates a significant degree of discomfort with the scales of forest landscape models.

2.4.6. Discomfort with Stochasticity and Variability in Simulated Processes

An important element in landscape ecological modeling is the stochasticity (whether random or biased), as well as spatial and temporal variability. We have found that users accustomed to deterministic knowledge may have difficulty dealing with probability and variability in simulated information. In our experience, users prefer the output of deterministic models, whether those outputs are a single numerical value or a map, and have difficulty with stochastic outputs such as probabilities and variances, whether depicted numerically or as choropleth maps. This is especially true when probabilities and variability are emergent properties of a simulation model and are not necessarily evident to the model's users in the input variables, model parameters, or model assumptions. At times, this discomfort appears peculiar or ironic to model developers because forest management professionals are well aware of the multiple and complex probabilities associated with ecological, economic, and social processes.

On the other hand, the discomfort of some users with stochasticity should not be surprising because it is not limited to users. Modelers also appear to prefer deterministic outputs. The number of deterministic simulation models far exceeds that of stochastic models. As well, model results are presented far more frequently as single deterministic values than as distributions of values, or as an expected value with surrounding probabilistic error. The general bias toward deterministic models is even more pronounced when considering landscape models and maps of model results. If users prefer deterministic model output, it is arguably because the modeling community has led them to expect it and be much less comfortable with probabilistic results.

2.4.7. Distrust of "Black Box" Models

Model users are occasionally uncomfortable using simulation models whose underlying mechanisms are not apparent. Usually, simulation models are designed to conceal complexities in structure and mechanisms, especially when they are designed for applied use. This may present a "black-boxed" appearance to users, who will not be confident in using a tool that they do not understand. Obscuring the model's logic can create significant barriers to use of the model, particularly when the policy or management context for use of the simulation model is contentious.

This distrust may also arise from failure to understand the model's assumptions. When these assumptions are obvious, users may be more tolerant and less distrustful of models that hide their mechanisms. We can be reasonably confident that few users want to see the numerical algorithms used to solve a model's differential equations or care whether the dynamics of a process are modeled using finite differences versus differential equations or ordinary versus partial differential equations or uniform square lattices versus vector-based mosaics of irregular polygons. But other assumptions are likely to be important. Which mechanisms should be made explicit and which should be hidden? Answering this question, and even distinguishing between a "mechanism" and an "assumption" (models do, after all, assume certain mechanisms) is more art than science, and there is no universal solution. Nevertheless, the necessity of hiding certain aspects of the model to enable its use by practitioners, and decisions about what to hide, will continue to be a barrier for certain users and uses.

At the same time, some users are too comfortable using black-box models or treat and use all models as if they were black boxes. As noted above, both scientists and practitioners often use models developed by others without careful consideration of the assumptions, methods, and implementation of the model, and how these factors might influence the results and their interpretation. The degree to which users expect, desire, and trust black-box models, and how this encourages or discourages their use of the models, varies widely.

2.4.8. Distrust of Methods of Model Validation

Forest management professionals are accustomed to forest ecological models for which empirical data can be readily obtained through observation or experimentation and used to validate the simulation results. Simulation models in landscape ecology, on the other hand, produce output that cannot be readily validated based on the user's experience or on available empirical data because of the breadth in scale, complexity, and stochasticity of the ecological processes being modeled. As a result, potential users may distrust these simulation models even when they reflect the best science available. This is especially evident with simulation models of long-term processes such as climate change, species migrations, and disturbance regimes. This is a valid criticism and a predictable impediment to the use of forest landscape ecological models. Especially when they are used to estimate or forecast quantities, and decisions are made to expend resources or enact legislation based on those results, users have difficulty knowing how much trust or confidence they can place in the results.

2.4.9. Unavailability of Computing Technology and Spatial Data

Almost all landscape ecological simulation models are spatially explicit, and require both large quantities of spatial data and significant computing capacity. The exceptional growth of desktop computing over the past 20 years, coincident with the growth of landscape ecology and having contributed significantly to growth of the discipline, has greatly reduced the technological barriers, but some users still may lack access to sufficient data or sufficient computing power to process the data. As well, modelers are always pushing the technological envelope and exploiting the latest computational capacity (e.g., high-performance parallel computing), and a gap will always exist between the needs of state-of-the-art models and the technology and data available to potential users of the models. If not managed properly, this gap can hinder widespread use of more advanced models.

2.4.10. Necessity for Third Party Involvement

The use of forest landscape ecological simulation models usually requires knowledge of geographical information systems, programming, spatial statistics, or all three disciplines. Since most users, whether policymakers or forest resource managers, have not learned this suite of skills during their formal training, they must often rely on experts who can use the models on their behalf, and these experts serve as translators of the messages being communicated by the model's developers. Since these technological experts may not necessarily have a landscape ecological, forest management, or policy development background, knowledge transfer becomes a three-way dialogue rather than a simple dialogue between developer and user. Although this dialogue has improved the application of models in some cases, we have also seen instances where the requirement for a third party acts as a barrier to acceptance of simulation models. We discussed this previously in our example of how scientists using models developed by other researchers and who attempt to transfer the models to decisionmakers or managers may themselves be unfamiliar with the model's assumptions and less sensitive to the model's limitations than those who developed the model. Thus, forest scientists themselves may be third parties and translators who become a barrier to appropriate use of a model.

2.5. IS MISUNDERSTANDING OF SCALE A SERIOUS IMPEDIMENT TO USERS OF LANDSCAPE ECOLOGY?

As noted in the introduction to this chapter, forest landscape ecology deals with large spatial extents with a spatial resolution that is generally much coarser than the individual trees or stands that are more familiar to practitioners. Changes in the forest landscape over these large spatial scales often take place only over long periods of time, and these slow dynamics can only be observed through sampling at relatively infrequent intervals. At the same time, our observational perspective is comparatively fine-grained. We observe daily, seasonal, and interannual changes in trees and stands that might be significant with respect to larger-scale changes, or might only be high-frequency "noise" (i.e., insignificant variation).

With scale so central to forest landscape ecology, misunderstanding of the importance of scale or a failure to incorporate principles of scale in modeling of forest landscapes is likely to create barriers to widespread use of forest landscape ecological models. Conversely, understanding of the importance of scale and disciplined treatment of scale could both provide potential solutions.

By and large, principles of scale are *not* having a positive impact on the application of landscape ecology to forest management. This assessment is based on our combined experience with principles of scale in ecological, landscape, and forest landscape modeling, and observations of their use in forest management. Others also have discussed how understanding the concepts of scale is important to forest landscape ecology and other applications of landscape ecology (e.g., Allen et al. 1984; Bissonette 1997). Forest resource managers are increasingly aware of the importance of scale as modern forest management is moving toward forest landscape management. Forest landscape models that are sensitive to issues of scale, and particularly to large and multiple scales, have been and are being developed with application in forest management as a primary goal. However, we believe that the full richness of the literature on the importance of scale in ecology has not been exploited in the development of models, and that an understanding of ecological scale is not informing the practice of forest landscape management.

This occurs for a variety of reasons. First and foremost, many practitioners simply do not see the relevance of concepts of scale beyond the idea that forested landscapes involve a large spatial extent. If their understanding of landscape ecology is limited to the notion that landscape ecology is simply the ecology of large areas or that landscapes are nothing more than large areas, they may feel they know all that they need to know about scale. Thus, the failure of landscape ecologists to emphasize aspects other than large spatial extents and to counter that bias may have created a barrier to practitioners pursuing a deeper understanding of scale.

Even practitioners and landscape scientists who have moved beyond that barrier may encounter additional barriers to understanding and incorporating the concepts of scale into modeling and practice. These include the possibilities that:

- There has been too little discussion of scale in ecology.
- The discussions of scale that have taken place may not have been sufficiently clear.
- The existing theory and principles of ecological scale may be too esoteric.
- Too much attention may have been placed on multiple scales rather than on the appropriate scale.
- Too little attention has been devoted to the defining principles of landscape ecology.
- There has been too little synthesis of our understanding of ecological scale.
- It is simply too soon for the science of ecological scales to significantly affect landscape management.

In the remainder of this section, we briefly address each of these issues.

2.5.1. Have We Discussed Enough about Scale?

Yes and no. Yes, because there has been much discussion of ecological scale in journal articles and books going back at least 20 years (e.g., O'Neill and King 1998). No, because the more important question is whether all that talk has been clear and effective in communicating the importance of scale to decisionmakers, managers, and other practitioners.

2.5.2. Has the Discussion of Scale Been Clear?

No, as a whole it has not been. There have been good discussions and explanations of the importance of scale in ecology (e.g., Turner et al. 1989), and individual presentations to potential users of this knowledge may have clearly and logically presented definitions of the concepts. However, the body of literature on scale often appears contradictory because different authors have investigated different problems, scales, or contexts without making these differences clear. This situation can generate confusion for individuals who are investigating how scale might influence an application. Perhaps more importantly, different individuals may develop different understandings of scale depending on which portion of the literature they sampled. Differences in understanding can lead to misunderstandings and confusion. The imprecise and inconsistent use of "scale" and "level" (as in the phrase "level of organization") is an example of one cause of confusion (Allen 1998; Allen and Hoekstra 1990; King 1997, 2005).

2.5.3. Has the Theory of Scale in Ecology Been too Esoteric?

Yes, at least in part. There are certainly commonsense aspects of scale that have influenced or are influencing forest modeling and management. One example is that large-scale systems such as forested landscapes require observations over large spatial extents and long time periods, and the scales of observation and management are increasingly being matched to the scale of the system. Similarly, there is an increasing recognition that the forest systems being managed encompass multiple scales, and new management approaches are addressing those different scales. There have also been largely theoretical discussions of scale explicitly targeted at an audience capable of applying this knowledge (e.g., Allen et al. 1984; King 1997). However, other aspects of the theory, including elements with rich potential insights on how to understand and manage large, multiscale systems, have tended to be couched in terms of unfamiliar abstractions and theoretical or mathematical terminology (O'Neill et al. 1989; Rosen 1989). The target audience for these presentations has been other scale researchers, which is fine so far as it goes, but the esoteric nature of these presentations, which comprise a sizable portion of the literature on scale in ecology, makes them unsuitable for practitioners. The differences in the language and style of presentation between researcher and practitioner audiences have been, in part, responsible for the limited influence of the discussion of scale in applied forest management. Some parts of the message are getting through; others are not.

2.5.4. Has There Been too Much Focus on Multiple Scales?

Yes. One of the recommendations to come from the consideration of scale in ecology has been a call for observations and studies at multiple scales. This is scientifically appropriate, but incomplete. It is certainly true that forested landscapes span a wide range of observational scales and involve processes operating at many different scales. It is also true that observations and studies at multiple scales will help determine how different processes operating at different scales are ultimately expressed at the scale of the forested landscape. But lost in this focus on multiple scales has been the equally fundamental message that there may be a single scale of observation, or a small set of scales, that is most appropriate to the specific management problem faced by a practitioner. If one has the objective of management of a forest at a given spatial extent for a given period of time, the theory of scale in ecology argues for finding *the* scale of observation most appropriate to that objective. It does not argue for, in fact argues *against*, looking at all scales encompassed by the scale of the management objective.

The principle of the appropriate scale for observing and understanding ecological systems draws heavily on hierarchy theory (Allen et al. 1984; King 1997, O'Neill 1989; Urban et al. 1987) and argues in favor of a three-scale approach. Hierarchy theory asserts that the focal level L of a system is the level of observable dynamics chosen by the investigator, and in the context of this chapter, is determined by the management objective. A mechanistic explanation of the dynamics at this level is found at the next lower level of organization $L-1$. However, level L occurs within the context of the next higher level $L+1$. This higher-level organization simultaneously bounds and is a consequence of focal level L, and both the constraints on the dynamics of that focal level and the significance or results of those dynamics can be found by examining the next higher level $L+1$. Allen et al. (1984) and King

(1997) and the references cited therein provide further details. Briefly, a three-level approach to nested, hierarchically organized ecological systems implies a corresponding three-scale approach to observing and understanding these ecological systems. The power of this approach lies in its emphasis on identifying the correct focal scale for a given management objective and the scales above and below that scale to discern the context and mechanisms (respectively) that govern that scale. It is this emphasis that has been lost in or obscured by the broader message that multiple scales are at work in any landscape.

A combination of hierarchy theory with the theory of scale can provide guidance on how to find the appropriate scales for a stated objective or application. This example illustrates how a richer understanding of scale can benefit applied forest landscape ecology. The consideration of scale in landscape ecology should be more nuanced than a simplistic recommendation to address only large scales or multiple scales. Such a message can be misinterpreted as a call for the study of multiple, arbitrarily selected scales even if those scales range from small to large. The arbitrary interpretation of multiple scales multiplies the problems for decisionmakers and managers, who are being asked to obtain scientifically sound observations and understanding at many different scales rather than at the most appropriate scale for their problem. The limited resources available to most practitioners would be better applied to identifying, observing, and understanding the most appropriate scale or limited number of scales for their management objectives.

Of course studies at multiple scales are needed to provide guidance for identifying the appropriate scales. Such studies might be required in circumstances in which theory provides uncertain or ambiguous guidance. Studies at multiple scales are also required in the determination of scaling rules or functions (King 1991; Milne 1997; Schneider 1994) that are used to translate information and observations across scales—for example, from the scales empirically accessible by field studies to larger scales of management objectives. In each case, there are uses for a fuller consideration of the importance of scale in landscape ecology.

2.5.5. Has There Been too Little Attention to the Defining Principles of Landscape Ecology?

Yes. Although this is not strictly an issue of scale, it is related to scale. Landscape ecology is a subdiscipline of ecology that focuses on understanding how spatial patterns and structures influence ecological processes (Turner 1989). It is true that most landscape ecology deals with spatial extents measured in thousands of hectares, but that tendency is historical and secondary, not a defining characteristic. The focus on how considerations of scale might help address a large spatial *extent* that encompasses processes at many different temporal and spatial scales has diverted attention from a consideration of how issues of scale might affect our understanding of spatial *patterns* and their influence on processes. Accordingly, the attention to scale, in the narrow sense of "large spatial scale," has detracted from the application of landscape ecology to forest management.

2.5.6. Has There Been Sufficient Synthesis?

No. The large and diverse literature on scale in ecology has not been sufficiently reviewed and synthesized from a scientific perspective. There has been even less effort devoted to synthesizing this knowledge from the perspective of potential application and to addressing problems using the language and examples familiar to the potential users of the knowledge. This lack of a useful and familiar synthesis has undoubtedly contributed to the limited application of considerations of scale and forest landscape ecology to forest management.

2.5.7. Perhaps It's too Early?

Perhaps. Intensive investigations of scale in ecology and the inevitable debates that have ensued go back more than 20 years. After that much time, one might hope for a more obvious influence of applications of scale in forest management and elsewhere than is currently apparent. The heightened awareness and understanding of issues of ecological scale in the scientific community has in fact influenced forest management to some degree; that is, the scientific deliberations on the challenges of large-scale ecological applications that influenced the growth of landscape ecology are gradually being transferred into applications. Today's discussions of forest management and ecological applications are different from those that occurred prior to the growth of landscape ecology and its considerations of scale. We suspect that the consideration of larger scales in modern forest management was driven primarily by the advent of satellite-based remote sensing and the accompanying changes in visual perspective, and that the emergence of landscape ecology was simultaneously influenced by these technological changes. But larger-scale applications and landscape ecology have grown together, have had positive influences on one another, and will likely continue to do so. Researchers and practitioners increasingly share their language, concepts, and understanding. The influence of science on practice undoubtedly requires more time to be fully realized. We may simply be anxious to see more impact and influence than the natural time scales of the feedback process permit. Nevertheless, the apparent influence has been patchy. Greater attention to the process of knowledge transfer to promote appropriate use of an understanding of scale in landscape modeling and forest management is called for.

2.6. HOW CAN RESEARCHERS MAKE SIMULATION MODELS MORE APPEALING TO USERS?

In this section, we offer some suggestions on how researchers who develop simulation models can more effectively transfer their scientific knowledge to users capable of applying that knowledge. Our intent is not to popularize simulation models, but rather to promote their judicious and appropriate use so that the gap between knowledge of forest landscape ecology and its application can be bridged in the long term. The points we address below are applicable whether the intended user of the knowledge is a forest resource manager who will use simulation models to support the development of tactical plans or a policymaker who will use simulation models for the development of strategic policy. Our discussion combines the views of several authors (e.g., Dale 2003a,b; Gutzwiller 2002b; Perera and Euler 2000; Turner et al. 2002) with the lessons learned from our own failures as developers of applied models.

2.6.1. Understand the User, Not Only the Use of the Models

When the goal of developers of landscape ecological models is the applied use of their models to solve problems, it is essential that they understand not just the intended use of the model but also the users and how this audience will use it. For example, the research community considers a model elegant if it embodies advanced scientific methods, logic, and computational techniques; in contrast, practitioners consider a model elegant if it is easy to use, appears simple and trustworthy, and produces useful and realistic results that support their efforts to solve problems. The elegant research model can be adapted so that it can be used to solve particular problems if it simulates the appropriate variables, at the appropriate temporal and spatial scales, in response to the appropriate drivers. However, that applicability alone may not be sufficient, and may even become counterproductive.

When a model requires too much data, is too computationally demanding, or involves state-of-the-art concepts and logic that are unfamiliar to the user (e.g., "fuzzy logic"), it may be applicable but it will not be applied. Knowing who the end users are—their educational background, geography, institutional and professional cultures, the resources they have available to implement models, and the practical difficulties they face—will help modelers to understand the users' perspective and develop models the users are likely to embrace. Accommodating user expectations by understanding who end users are and their specific needs does not diminish and compromise the scientific rigor of a simulation model; rather, it enhances its appeal to the users.

2.6.2. Develop Simple Models

Simplicity in model development is a desirable quality that will increase user acceptance as well as the model's ease of use. To many model developers, simplifying models means little more than adding a graphical user interface (GUI) between the model and the user. GUIs are useful for concealing intricacies that are not necessary for use of the model, and thereby increase the perceived user-friendliness and convenience of the model. However, models can be made even easier to use and more attractive to the user by designing them based on the principle of parsimony. Potential users of a model may demand more details than are necessary, and in this case, the developer may need to emphasize simplicity over the complexity that would result from addressing all their demands. By parsimony, we mean that the goal is to reduce the complexity of the model's mechanics by judicious choice of the model's scale, functions, parameters, input data, and outputs. What we suggest goes beyond the typical sensitivity analyses carried out during model construction, in which the model outputs provide the sole guidance. We urge modelers to define the most parsimonious model possible given the user's requirements. Developers should mine the literature on scale in ecological systems for insight into how mechanisms and their functional representation vary with scale in scale-dependent levels of organization. Understanding how the scale of observation (the observer's perspective) influences how the system looks to the observer provides insight into how to adapt the appearance of the model to the user. Another suggestion is to consider developing multilayered models, which contain a hierarchy of submodels that can be coupled or decoupled to attain levels of complexity that can be tailored to the needs of each user. The points at which coupling and decoupling can occur might coincide with scale-dependent levels of organization in a hierarchically organized system, or might reflect how users analyze and interact with the components of the problem.

Designing for simplicity based on considerations of scale and the user's perspective will also help designers to determine which mechanisms should be placed in black boxes (i.e., made invisible to the user of the model). For example, fine-scale mechanisms that are far removed from the larger scale of observation can be concealed so that only aspects (e.g., aggregate properties) that are translated across intervening scales will be presented to the user. These are the kinds of design decisions that are intuitively made while designing for parsimony and simplicity. For example, a deeper understanding of scale in ecological systems could be used to develop simpler models that are parsimonious with respect to scale. The three-level models suggested by the hierarchy theory discussed in Section 5.4 provide one example of this approach. King (1997) discusses application of this approach to an age-structured population model.

2.6.3. Clarify the Limitations of the Model to Its Users

Simulation models are applicable under very specific conditions and assumptions, and are only suitable for specific uses. Model developers cannot assume that these limitations, assumptions, and objectives are clear to the model's users. As we mentioned earlier, ambiguity in explaining a model's limitations leads to misuse in the short term and mistrust in landscape ecological applications in the long term. Researchers must thus make a concerted effort to clearly articulate the intended use (e.g., exploratory versus forecasting) of their models. For example, simulation models developed for discovery and exploration are useful tools to provide contextual information such as the consequences of climate change. However, such models must not be used to forecast specific scenarios, however convincing they may be, or to help managers make tactical management decisions as though they were decision-support systems. Another example relates to simulation models of historical landscapes. Although some of these models provide insights into how present landscapes may have evolved, and into landscape patterns and processes in a different temporal context, direct use of such models to generate blueprints of future landscapes is questionable.

Strict assumptions are fundamental to developing good models, but violating these assumptions (e.g., changing scales, intervals, extents, and periods of application or modifying state variables) can generate false outputs and incorrect inferences. For example, a probabilistic simulation model of the incidence of insect epidemics developed for the current climate and forest composition must not be advocated for long-term use because climate and forest composition may not be static over longer periods. Neither is that model suitable for deterministic spatiotemporal forecasts of the incidence of epidemics. Model developers must ensure that the assumptions and limitations in terms of scale, resolution, ranges of the state variables, and model functions are clear to users, and that users understand the consequences of violating those conditions.

Designing and developing models with simplicity as an objective facilitates the communication of assumptions, limitations, and consequences of violating these premises to the user. It is difficult to understand and communicate all of the assumptions—or even the most critical assumptions—in a complicated forest landscape simulation model. A more parsimonious model is easier for both the developer and the user to understand, and it is easier to explicitly communicate the assumptions and limits of simpler models.

In general, users require more explicit communication of the purpose of a model and the degree of confidence they should place in its outputs, and model results should be presented as probabilities with associated confidence intervals. However, the modeling community should also invest more effort in establishing methods and protocols for determining and communicating how much confidence should be placed in the outputs of their models. As noted above, traditional validation of models against observations is frequently impossible because of the scales involved. Accordingly, alternative approaches for evaluating model performance must be established and communicated to users. Because this is an important consideration that we cannot fully address here, we refer readers to discussions of nontraditional methods for testing model predictions (e.g., Gardner and Urban 2003; Kleindorfer et al. 1998; Oreskes 1998, 2004; Oreskes et al. 1994; Sargent 2004).

2.6.4. Transfer Knowledge to Users Interactively, before, during, and after Model Development

Knowledge transfer related to forest landscape ecological simulation models must extend beyond passive means such as publications, posters, and oral presentations, especially when there is a clear group of users for a model. Whenever possible, researchers must actively initiate and engage in knowledge transfer to users of the model and, rather than waiting to begin knowledge transfer until after the models have been developed, should strive to initiate a dialogue between developers and users at the design stage and continue this dialogue through development and testing of the model. The most effective and appropriately used models will be those that are designed and modified to meet specific user needs, following explicit definition of specifications by the users and iterative improvement based on feedback from the users.

Model developers can engage end users in many ways. Here, we suggest only a few possible avenues. At the outset of the design stage, researchers should engage in a dialogue with the intended users of their model to understand their specific needs and the context in which those needs will be met, and to convey the concepts underlying the model. This exchange is vital because understanding of these concepts is essential to prevent users from subsequently using the model in an inappropriate context. At this stage, users and model developers can also establish a shared vocabulary to prevent miscommunication later in the process. Adopting and adapting the use of formal model specifications would prove useful.

As model design progresses, developers can inform users of the logic and principles behind the model to ensure that they understand and accept the modeling methods. An early understanding and acceptance of model logic and methods by users is preferable to basing eventual acceptance solely on validating the model results using empirical data—something that may not even be possible for some types of model. Many modeling approaches can yield similar matches with observations, but different users will prefer or require different approaches for different uses. Based on this continuing dialogue, model developers will increasingly understand design calibrations that are necessary for the model to meet the needs of its users, and users will increasingly understand the limitations and assumptions that govern use of the model.

Postdevelopment model testing and sensitivity analyses should be conducted using user-provided data, creating another opportunity for users to understand the model and provide feedback to developers. Every interaction between developers and users of the model can be a mutually productive knowledge transfer opportunity and learning experience.

2.6.5. Synthesize an Understanding of Scale from the Perspective of Model Application

Our recommendations for making simulation models more appealing to their users should be complemented by a comprehensive review and synthesis of the current understanding of scale in ecology, and in landscape ecology in particular. The specific aim of this synthesis should be to make what is currently known about scale in ecology more useful in the realm of application. The synthesis should use language and examples familiar to practitioners and other users of the model, and be designed to be used during the process of developing and deploying the model.

2.7. MUTUAL BENEFITS

In this chapter, we have focused on challenges to the transfer and extension of forest landscape modeling knowledge into forest management, and on possible solutions. Our premise has been that forest management will benefit from appropriate use of forest landscape models, and that this use will benefit from explicit consideration of the users and their needs during design and deployment of the models. However, the benefits are mutual. Obviously, model developers who intend for their model to be used in practical applications will benefit when the model is actually used and is used appropriately. But, their discriminate use will also benefit when forest landscape models are designed with these considerations in mind.

It can be argued that one of the guiding principles of Western scientific endeavor is the desire to explain a complex natural world using a finite and relatively small set of simple relationships or laws. Science seeks explanations through simplification and parsimony, and the principle of parsimonious model design that we have proposed in this chapter is in keeping with that goal. We propose that a more explicit and formal consideration of the principles of ecological scale will help move the design of forest landscape ecological models from art to science. Forest management will benefit from better-designed forest landscape models, as will the science of forest landscape ecology.

LITERATURE CITED

Allen TFH (1998) The landscape "level" is dead: persuading the family to take it off the respirator. In: Peterson DL, Parker VT (eds) Ecological scale: theory and applications. Columbia Univ Press, New York, pp 35–54.

Allen TFH, Hoekstra TW (1990) The confusion between scale-defined levels and conventional levels of organization in ecology. J Veg Sci 1:5–12.

Allen TFH, O'Neill RV, Hoekstra TW (1984) Interlevel relationships in ecological research and management: some working principles from hierarchy theory. USDA For Serv, Rocky Mountain For and Range Exp Stn, Fort Collins, Colorado. Gen Tech Rep RM-110.

Bissonette, JA (1997) Wildlife and landscape ecology: effects of pattern and scale. Springer, New York.

Buongiorno J, Gilless JK (2003) Decision methods for forest resource management. Elsevier Science, San Diego.

Dale VH (2003a) Opportunities for using ecological models for resource management. In: Dale VH (ed) Ecological modeling for resource management. Springer-Verlag, New York, pp 3–19.

Dale VH (2003b) New directions in ecological modeling for resource management. In: Dale VH (ed) Ecological modeling for resource management. Springer-Verlag, New York, pp 310–320.

Gardner RH, Urban DL (2003) Model validation and testing: past lessons, present concerns, future prospects. In: Canham CD, Cole JJ, Laurenroth WK (eds) Models of ecosystem in science. Princeton Univ Press, Princeton, pp 186–205.

Gutzwiller KJ (2002a) Applying landscape ecology in biological conservation: principles, constraints, and prospects. In: Gutzwiller KJ (ed). Applying landscape ecology in biological conservation. Springer-Verlag, New York, pp 481–495.

Gutzwiller KJ (ed) (2002b) Applying landscape ecology in biological conservation. Springer-Verlag, New York.

Jansen M, Judas M, Saborowski J (eds) (2002) Spatial modeling in forest ecology and management: a case study. Springer-Verlag, New York.

King AW (1991) Translating models across scales in the landscape. In: Turner MG, Gardner RH (eds) Quantitative methods in landscape ecology: the analysis and interpretation of landscape heterogeneity. Springer-Verlag, New York, pp 479–517.

King AW (1997) Hierarchy theory: a guide to system structure. In: Bissonette JA (ed) Wildlife and landscape ecology: effects of pattern and scale. Springer-Verlag, New York, pp 185–212.

King AW (2005) Hierarchy theory and the landscape... level? or Words do matter. In: Wiens J, Moss M (eds) Issues and perspectives in landscape ecology, Cambridge Univ Press, Cambridge, UK, pp 29–35.

Kleindorfer GB, O'Neill L, Gansehan R (1998) Validation and simulation: various positions in the philosophy of science. Manage Sci 44(8):1087–1099.

Milne BT (1997) Applications of fractal geometry in wildlife biology. In: Bissonette JA (ed) Wildlife and landscape ecology: effects of pattern and scale. Springer-Verlag, New York, pp 32–69.

Mladenoff DJ, Baker WL (eds) (1999) Advances in spatial modeling of forest landscape change: approaches and applications. Cambridge Univ Press, Cambridge, UK.

O'Neill, RV (1989) Perspectives in hierarchy and scale. In: Roughgarden J, May RM, Levin SA (eds) Perspectives in ecological theory. Princeton Univ Press, Princeton, pp 140–156.

O'Neill RV, King AW (1998) Homage to St. Michael; or, why are there so many books on scale? In: Peterson DL, Parker VT (eds) Ecological scale: theory and applications. Columbia Univ Press, New York, pp 3–15.

O'Neill RV, Johnson AR, King AW (1989) A hierarchical framework for the analysis of scale. Landscape Ecol 3:193–206.

Oreskes N (1998) Evaluation (not validation) of quantitative models. Environ Health Perspect 106 (Suppl 6):1453–1460.

Oreskes N (2004) Science and public policy: what's proof got to do with it? Environ Sci Policy 7:369–383.

Oreskes N, Shrader-Frechette K, Belitz K (1994) Verification, validation, and confirmation of numerical models in the earth sciences. Science 263:641–646.

Perera AH, Euler DL (2000) Landscape ecology in forest management: an introduction. In: Perera AH, Euler DL, Thompson ID (eds) Ecology of a managed terrestrial landscape: patterns and processes of forest landscapes in Ontario. Univ British Columbia Press, Vancouver, pp 3–12.

Perera AH, Buse LJ, Weber M (eds) (2004) Emulating forest landscape disturbances: concepts and applications. Columbia Univ Press, New York.

Perera AH, Buse LJ, Crow TR (2006) Knowledge transfer in forest landscape ecology: a primer. In: Perera AH, Buse LJ, Crow TR (eds) Forest landscape ecology: transferring knowledge to practice. Springer, New York.

Rosen R (1989) Similitude, similarity, and scaling. Landscape Ecol 3:207–216.

Sargent RG (2004) Validation and verification of simulation models. In: Ingalls RG, Rossetti MD, Smith JS, Peters BA (eds) Proceedings of Winter Simulation Conference. Inst Electrical and Electronics Engineers, Piscataway, New Jersey, pp 17–28.

Schneider DC (1994) Quantitative ecology: spatial and temporal scaling. Academic Press, San Diego.

Turner MG (1989) Landscape ecology: the effect of pattern on process. Annu Rev Ecol Syst 20:171–197.

Turner MG, O'Neill RV, Gardner RH, Milne BT (1989) Effects of changing spatial scale on the analysis of landscape pattern. Landscape Ecol 3:153–162.

Turner MG, Crow TR, Liu J, Rabe D, Rabeni CF, Soranno PA, Taylor WW, Vogt KA, Wiens JA (2002) Bridging the gap between landscape ecology and natural resource management. In: Liu J, Taylor WW (eds) Integrating landscape ecology into natural resource management. Cambridge Univ Press, Cambridge, UK, pp 433–465.

Urban DL, O'Neill RV, Shugart HH Jr (1987) Landscape ecology. BioScience 37:119–127.

3

A Collaborative, Iterative Approach to Transferring Modeling Technology to Land Managers

Eric J. Gustafson, Brian R. Sturtevant, and Andrew Fall

3.1. Introduction
3.2. Conceptual Framework of the Technology Transfer Process
3.3. Description of the Modeling Technology to Be Transferred
    3.3.1. SELES
    3.3.2. LANDIS
3.4. Case Study 1: The Morice Land and Resource Management Plan
    3.4.1. Overview of the Morice Landscape Model
    3.4.2. Process Models
    3.4.3. Indicator Models
    3.4.4. Outcomes
    3.4.5. Obstacles and Lessons Learned
3.5. Case Study 2: Managing Fire Risk in the Wildland–Urban Interface
    3.5.1. Outcomes
    3.5.2. Obstacles and Lessons Learned
3.6. Comparison of the Case Studies
3.7. Conclusions
    Literature Cited

ERIC J. GUSTAFSON and BRIAN R. STURTEVANT • USDA Forest Service, North Central Research Station, 5985 Highway K, Rhinelander, WI 54501, USA.   ANDREW FALL • Simon Fraser University, 8888 University Drive, Burnaby, BC V5A 1S6, Canada.

3.1. INTRODUCTION

Land managers have come to realize that achieving many natural resource management goals requires a consideration of landscape-level patterns and processes (Boutin and Hebert 2002). A landscape perspective is necessary because many ecological processes operate at landscape scales (Turner 1989). For example, although silvicultural techniques applied at the stand level can be used to manipulate species composition and growth form, there are landscape-level processes (e.g., wind, fire, insect outbreaks) that also have significant effects on these stand characteristics (Liu and Ashton 2004). In a reciprocal way, landscape patterns also determine the likelihood of insect and disease outbreaks (Sturtevant et al. 2004a), fire ignition and spread (Hargrove et al. 2000), browsing by deer (Alverson et al. 1988), the influx of invasive species (With 2002), and pollution (Weathers et al. 2001). Each of these landscape-level ecological processes can influence the achievement of local-scale management objectives. Unfortunately, these multiscale processes sometimes interact in complex ways that are difficult to predict (Turner et al. 1994). Land managers therefore depend on sophisticated technology to predict the consequences of proposed management actions.

Public land-management agencies are particularly well positioned to implement landscape-level management because of the size of the land base under their jurisdiction. However, the complexities of ecological phenomena across scales and of the relationships among ecological processes are particularly daunting given the public's low tolerance for management mistakes (Teich et al. 2004). Therefore, public land-management agencies require analytical and prediction tools for modeling landscape processes and evaluating strategic management options in the context of changing management environments. These modeling tools must adequately represent the ecological system and the alternative management options. The representation of the ecological system must fully account for the interactions among all relevant ecological processes and management activities. For example, the management of fire risk can involve reducing the fuel loads in forest stands (either by manual removal of biomass or by prescribed burning), manipulating forest composition to encourage less-flammable species, manipulating the spatial pattern of the landscape mosaic to reduce the likelihood of fire spread, or taking action to control ignitions (e.g., road closures, campfire prohibitions). However, these actions may also affect forest succession, or the likelihood of insect pest outbreaks or catastrophic blowdowns, all of which affect the risk of fire (Fleming et al. 2002). Models that allow an assessment of fire risk under various alternative management options while accounting for these potential interactions should result in better decisions than when stand-level or overly simplistic models are the only source of support.

A variety of landscape-level analytical and projection models have been developed for various research and management purposes (Baker 1989; Sklar and Costanza 1991). In some cases, these models can be used to directly answer specific management questions and support decisions. In other cases, new models must be developed to provide specific information that is lacking from other models. Transferring these models or the information they generate from the developers to

A Collaborative, Iterative Approach

the managers who will use these resources is usually difficult. In this chapter, we explore the reasons for this difficulty, and outline a collaborative, iterative approach to the transfer of modeling technology. We describe two case studies that illustrate the approach for specific management problems. We conclude by discussing the merits of the approach and the lessons learned through the case studies.

3.2. CONCEPTUAL FRAMEWORK OF THE TECHNOLOGY TRANSFER PROCESS

Technology transfer has often been approached as a marketing problem. Researchers view the technology they have developed as a product that they must "sell" to potential users, who are often referred to as "customers" or "clients." This paradigm sets up a buyer–seller mentality that can hinder the successful adoption of complex technology by managers. We believe that the transfer of complex decision-support models presents at least seven formidable difficulties:

- Teaching managers or their support staff to run modeling software requires formal training and technical support. This is not unlike the process of learning any new software, but in addition, it requires an explanation of basic modeling concepts.
- Proper application of models by managers requires that they understand in some detail the assumptions behind the models and the limitations of the results. There is no small danger that inappropriate conclusions can be drawn should users of the models misunderstand the key underlying assumptions.
- Managers must learn how to interpret a model's results to provide defensible support for their decisions.
- Managers must precisely explain to researchers the decisions they must make and the information or knowledge required to make those decisions. Such an understanding will help researchers to judge whether the model can in fact produce the information that is needed.
- Political, funding, or logistical limitations may constrain management options. These issues may not be apparent to researchers, so managers must identify them; researchers could also interview the managers to identify any relevant constraints.
- Researchers may be unfamiliar with the specific land base that is being managed, and may therefore be poorly equipped to accurately model the ecological or management dynamics.
- Managers and researchers may have different understandings of uncertainty and of the risks associated with that uncertainty. A shared understanding of the role of uncertainty in the decisionmaking process is critical.

A common thread among these difficulties is the necessity for substantive communication and partnership between researchers and managers.

To resolve these difficulties, we adopted a collaborative, iterative approach for technology transfer (Ahern 2002, Fall et al. 2001). The approach is *collaborative* because it assembles a triad of researchers, management planners, and local resource experts. It is *iterative* because communication among the triad partners must occur repeatedly, so that the application of the tool can be refined with each iteration. Our approach fosters a "community of practice" in which people build understanding together in a social, physical, and temporal setting (Allee 1997, p. 219).

The conceptual framework for our collaborative, iterative approach to technology transfer is best represented as a triangular interaction among researchers, management planners and decisionmakers, and local resource experts (Figure 3.1). The interaction takes the form of iterative communication (the arrows in Figure 3.1), and the focus of the communication is on application of the modeling technology to support a particular management decision. Thus, the modeling tools serve as a common framework that lets all parties conceptualize and formalize (i.e., model) the management problem. The end result is a more defensible decision and a transfer of modeling technology from a research environment to a management environment. This approach also fosters the development of a shared vision of the model's requirements and the decision process to be supported, of the data requirements, of the interactions among ecological and human processes, of the model's capabilities and assumptions, and of how to appropriately interpret the model's outputs.

The collaborative nature of the process is important because each partner provides expertise that is critical to successful technology transfer. The researchers understand the feasibilities of applying existing models or building new models, know the assumptions that underlie a model, are familiar with the algorithms that drive a model, know how to estimate the model's parameters and develop the input data, provide the technical expertise to run a model, and guide interpretation of the

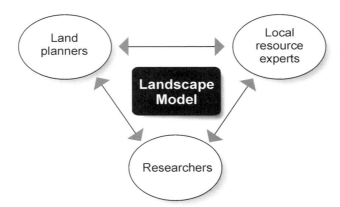

Figure 3.1. A conceptual framework for the collaborative, iterative approach to technology transfer. A triad of decisionmakers, researchers, and local experts collaborate to implement modeling technology and provide decision support. The arrows represent iterative communication.

model's results. Models are not a mysterious black box to the researchers who developed them, and this knowledge helps them to make the models more transparent to the other partners, giving them greater confidence in the output. The management planners understand the management decision that must be made, and can readily identify the information gaps that hamper their ability to make defensible decisions. They also can identify the bounds of politically or logistically feasible alternatives. Without such input, researchers are likely to develop elegant and sophisticated answers to irrelevant questions. The local resource experts enable the model application to reflect the best current knowledge about the system under study. They help the researchers to estimate realistic values of model parameters for the local ecosystems. They can readily identify model behaviors that incorrectly simulate the local reality and assist the planners to develop ecologically feasible management options. A two-way collaboration that only includes the researchers and planners is more likely to result in biologically indefensible results.

The collaborative interaction begins with a meeting of all three groups. The initial iteration focuses on sharing of information about the management decision to be made, the alternatives that will be compared, data availability, and the modeling tool to be applied. Resource experts inject biological reality into the discussion. Following this meeting, the researchers design the modeling protocols, work with the resource experts to estimate the model parameters, oversee the generation of input data, and perform the initial simulation runs. During this time, the researchers contact resource experts to clarify and refine the initial parameters. The second iteration brings all parties together again to review the initial model outputs. The resource experts assess whether the model's behavior is consistent with their understanding of the ecological system. The planners assess whether the information generated by the model is what they need to make a decision, and if not, work with the researchers to refine the modeling objectives. The researchers communicate any needs for better parameter estimates or additional information about management alternatives. A second round of simulations is then conducted based on the improved understanding of the management problem. The process is repeated until all parties are satisfied that the model is producing the information required to make the decision. Collaboration and iteration produce several important outcomes:

- The model results are of greater quality and relevance to the decisionmaker.
- Managers learn to use a new technology.
- Researchers learn about management problems and the constraints that managers face.
- Resource experts come to better appreciate the interactions among many resources and the realities of multiple-use planning.

The collaborative nature of the approach provides the synergy required for effective technology transfer.

To illustrate the application and utility of the collaborative, iterative approach to technology transfer, we will describe two case studies in which modeling technologies were successfully transferred from a research and development environment to operational use, to meet a management decision-support need. The second case study is a technology-transfer effort in progress. The first used the SELES modeling language to construct a new model that would support a complex and controversial land-use planning process. The second used the LANDIS model to predict how patterns of human settlement and forest management in the Nicolet National Forest (Wisconsin, USA) might intersect to influence fire risk. Before presenting the case studies, we have provided a brief orientation to the modeling technologies used and the philosophy behind their development to describe the basic principles of the underlying science.

3.3. DESCRIPTION OF THE MODELING TECHNOLOGY TO BE TRANSFERRED

3.3.1. SELES

SELES (the Spatially Explicit Landscape Event Simulator) is a model-building and simulation environment that attempts to strike a balance between the flexibility of a programming language that can be used to construct novel models and the ease of applying and parameterizing (estimating the parameters for) existing models (Fall and Fall 2001). Its foundation is a declarative language that lets its users focus on defining the specific needs of landscape modeling and analysis rather than a strictly procedural language that focuses on the details of computer program execution. By providing a language closely adapted to landscape ecology and spatiotemporal modeling, it allows relatively rapid and transparent development of models. For example, models can be written in the form of text files that are loaded directly into SELES rather than imposing the complexity of a conventional software development environment. The underlying assumptions are therefore explicit and not hidden within a black-box program that cannot be examined by anyone other than the programmer, and are not inseparably intertwined with the complicated programming code that performs the actual simulation. SELES models have been successfully applied to support landscape-level forestry decision processes in land-use planning (Fall et al. 2004a; Morgan et al. 2002), the management of natural disturbance (Fall et al. 2004b), and parks planning (Manseau et al. 2002).

The model-development process can be conceptualized using the analogy of car production (Figure 3.2). The overall objectives for a new car (by analogy, the landscape model) are set by marketers and clients (stakeholders). The design is created by engineers (modelers and resource experts) who understand automobile (modeling) capabilities, and the constraints and opportunities imposed by materials and aerodynamics (system knowledge and feasibility thresholds). They combine experience and knowledge with design objectives to create a blueprint (a conceptual model). It is important to note that the designers do not actually build the car, so they do not need to know the details of the implementation tools and technology, but they do need to understand the potential capabilities.

A Collaborative, Iterative Approach

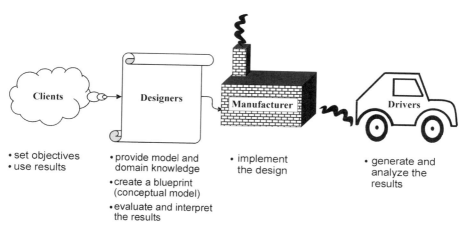

Figure 3.2. The various people involved in the modeling process, starting with those who set overall project goals and ending with those who analyze the model's outputs. The drivers are linked iteratively with the clients to communicate results and ensure that the desired outcomes are achieved.

The people who construct the car (build the model) must understand the operation of the factory machines and input resources (model-building tools and input data). They must be capable of understanding a design blueprint and implementing it (starting with prototypes), but do not necessarily need to be able to create designs. They must be able to test the car (the model) by means of test drives to ensure that its implementation matches the blueprint (preliminary model verification). This leads to the first level of iteration, which is performed entirely within the factory by the manufacturer (the modeling research team). Given a product that matches its design specifications, further testing is then required to assess how well its performance matches the design objectives. This may lead to a second level of iteration, indicated by the feedback arrow between designers and developers in Figure 3.1, in which the designers refine any of the blueprints that require updated implementations.

Given a car that matches its specifications (a verified model), a driver (the user of the model) can take the car for a ride (conduct a simulation). The driver does not need to understand the details of the blueprint or the manufacturing process. However, a user guide describing the control features and the consequences of manipulation of the controls is essential, as is an appropriate road map and destination (directions for what scenarios to assess with the model). Some basic knowledge of mechanics (understanding of basic information on modeling) can help the driver (the model user) make minor repairs or modifications and customize the vehicle (the model) in order to make the fullest use of the product and to reduce reliance on the factory workers. Understanding the assumptions that govern the use of the car (the model), such as the fact that the car is not designed to operate in a body of water, is also necessary. For something as familiar as a car, this knowledge is implicit, but for modeling, the knowledge is more abstract and must be taught to the model's users. Dials and other feedback features (progress indicators in the modeling software) provide indicators during a drive and a travel log (model output indicators) provides a long-term record.

Assessing whether the car meets its original objectives will likely require a number of drives under different conditions. Summaries of the travel logs can be analyzed by designers and presented to those who defined the original objectives (stakeholders), completing the third level of iteration.

The key features of this analogy that are relevant to the development of a model are:

- The process is inherently collaborative, with multiple levels of feedback and iteration. Products at one stage are used to refine the objectives (higher-level specifications) for subsequent stages. The feedback emerges through communication between all parties during all stages to ensure that objectives are met and that later stages encapsulate the results of earlier stages.
- No one person performs all modeling tasks. A range of modeling expertise is required for success, particularly for model design (abstraction of reality), model implementation and testing, and model utilization and analysis.
- The process involves multiple levels of abstraction, and people at one level in the process only need to fully understand the aspects of the overall system that are relevant to their level. This helps to reduce the scope of the problems to be addressed during any given stage, and minimizes the level of perceived complexity at any stage.

Modeling tools can be placed along a spectrum from traditional procedural programming languages to fully constructed models (Fall and Fall 2001). Building models directly by programming is akin to constructing a new car without a specialized factory, using generic hardware and tools. On the other hand, using a prebuilt model is akin to using the same car for all types of transportation needs (i.e., fitting the question to the model rather than vice versa). SELES fills a niche by providing a "model-building factory" that facilitates the production of customized models that suit the user's objectives.

3.3.2. LANDIS

LANDIS is a landscape model that simulates spatial forest dynamics, including forest succession, seed dispersal, species establishment, various disturbances, and the interactions among these factors (Gustafson et al. 2000; Mladenoff and He 1999). LANDIS was developed as a research tool to simulate the reciprocal effects of disturbance processes (i.e., fire, wind, vegetation management, insect outbreaks) on patterns of forest vegetation and vice versa across landscapes up to 1 million ha in size and long time scales (50 to 1000 years). Our purpose here is not to describe LANDIS in great detail, but rather to demonstrate the model's power and the complexity of parameterizing and using the model.

LANDIS (v 4.0) uses a raster-based map (i.e., a grid), in which each cell contains information on the presence or absence of a tree species and the 10-year age cohort of that species, but no information about the number or size of the individual stems. Forest succession processes are simulated based on the relative ages of the species found in each cell and on the vital attributes of the species (e.g., shade tolerance, probability of establishment within each land type, longevity, seed dispersal distance). Disturbances remove certain age classes, and the number and characteristics of these classes depend on the severity of the disturbance, which in turn is determined by the characteristics of each cell and in some cases by the characteristics of nearby cells. The number of occurrences and spatial extent of a disturbance are determined by a number of parameters that typically vary as a function of land type. For example, the fire regime for a given land type is defined by the mean size of fires, fuel accumulation rate, and fire-return interval, which is defined as the average number of years required to burn an area equal to the area of the land type on the landscape. Forest management activities are specified by a spatial component (algorithms and spatial zones that determine the order in which stands are selected for treatment), a temporal component that specifies the timing of treatments, and a removal component that specifies which age classes are removed by the treatment. Readers interested in more details of the model should consult Gustafson et al. (2000), He et al. (1999a,b, 2004), He and Mladenoff (1999), Mladenoff and He (1999), Sturtevant et al. (2004a), and Yang et al. (2004).

A powerful attribute of this modeling approach is the feedback between disturbance and species response. For example, windthrow events may alter the species composition relative to sites without windthrow, and will contribute to fuel accumulation on a site, increasing the severity of subsequent fire events. The forest harvest module of LANDIS provides the ability to simulate specific and complex management alternatives, including timber extraction, fuel reduction treatments, and prescribed fires. The interaction of such treatments with natural disturbances and with the dynamics of forest succession produces powerful insights into the complex cumulative effects of specific proposed actions. For example, LANDIS simulations have shown an interaction between ecological land type (defined by land form, soils, and climate), management prescriptions, and initial conditions, and have predicted a highly variable risk of canopy fire in northern Wisconsin, USA (Gustafson et al. 2004; Sturtevant et al. 2004b).

Although LANDIS is a powerful projection tool, it is also quite complicated to use. The model requires input maps that specify the initial forest conditions for each cell of the grid. Preparation of these maps requires sophisticated statistical estimation and mapping techniques (He and Mladenoff 1999). Depending on the species, at least eight vital attributes must be parameterized for each species, and scaling methods are typically used to estimate species establishment coefficients for each land type (He et al. 1999b). At least 15 parameters are required to define the statistical distributions of natural disturbance regimes and fuel accumulation rates for each land type. A new fuel module (He et al. 2004) requires three times as many parameters as previous versions of the model. The sheer volume of the model outputs and the computational demands of the simulations require a fairly powerful desktop computer. For these reasons, LANDIS applications are somewhat intimidating for most land managers.

3.4. CASE STUDY 1: THE MORICE LAND AND RESOURCE MANAGEMENT PLAN

The northwestern interior area of British Columbia, Canada, is rich in biodiversity and in natural resource values (e.g., forestry, mining, tourism), and its landscape has a complex and varied geography (e.g., plateaus, fiordlike lakes, glaciated mountains). Different value systems (e.g., conservation versus resource development) have created conflicting perspectives on the best land uses. To help resolve conflicts and guide future land use in this area, the provincial government initiated a multistakeholder land-use planning process to develop the Morice Land and Resource Management Plan, which covers an area of approximately 1.5 million ha. The goal of the planning process was to reach agreement on land-use zoning (e.g., protected areas versus intensive or general management) and objectives (e.g., a sustainable and viable forestry sector and conservation of threatened species).

Assessing the risks to ecological values and the economic opportunities in this area required an analysis of complex spatial and temporal interactions among key landscape processes and states. A government technical team was formed to provide decision support to the Land and Resource Management Plan planning group (stakeholder representatives) by capturing knowledge and distributing information to support the planning process. These groups were both part of the "land planners" category in the conceptual framework shown in Figure 3.1. This decision-support system combined projections of a variety of processes and indicators to create an integrated landscape-analysis system, and at its core was a spatial landscape dynamics model, the Morice Landscape Model (MLM). The MLM was constructed using the collaborative, iterative landscape-analysis framework (Fall et al. 2001) by adapting models from prior projects (e.g., Morgan et al. 2002), and was implemented using the SELES (Fall and Fall 2001) modeling tool. A core modeling team worked with local resource experts and the government technical team, who in turn worked with the Land and Resource Management Plan planning group to communicate the implications of alternative scenarios and project the impacts of landscape change on timber supply, biodiversity, and species of concern.

The critical processes that were modeled included forest growth, natural disturbance, harvesting, and road construction. Resource experts conducted postsimulation interpretation and analysis of the MLM results to determine the effects of each scenario on species representative of healthy ecosystems, including the grizzly bear (*Ursus horribilis*), mountain goat (*Oreamnos americanus*), northern goshawk (*Accipiter gentilis*), and woodland caribou (*Rangifer tarandus caribou*). Because the MLM simulates both economic and ecological processes, it was possible to identify trade-offs between economic activities and ecological risk, and to define quantitative boundaries to the social trade-offs among the values for this landscape that were emphasized by the various stakeholders.

For successful technology transfer, it was critical for participants to recognize that the MLM was embedded in a human network of resource experts, special interest groups, stakeholders, and decisionmakers. The landscape model provided a tool that allowed experts to explore the decision space and to assess the existing and potential management regimes by evaluating indicators of the ecosystem's state, conducting experiments, and defining the bounds of the problem (i.e., identifying the feasibility limits for solutions to the problem). Resource experts gained an understanding of how the landscape, wildlife, and vegetation would change given certain human interventions and natural processes. Through the government technical team, this information was then distilled and communicated to the Land and Resource Management Plan planning group (which comprised people with and without technical expertise) in a form that helped them converge on a final plan. The critical point was that the transfer of analytical and technological information from the model occurred via the interaction between the resource experts and the planning group, and it was this human component of the system that explained the knowledge and information derived from the model to the decisionmakers. Hence, during the transfer process, information was transformed from highly quantitative and detailed data (e.g., geospatial inputs, yield tables, constraints, zones) into more qualitative and general data regarding ecological and economic risks.

3.4.1. Overview of the Morice Landscape Model

This section briefly describes the main concepts and assumptions underlying the MLM. The definition of this model using SELES consists of a linked set of two types of submodel: submodels of landscape change and submodels that calculate indicators for timber supply and ecological risk. The inputs consisted of digital raster maps at a 1-ha resolution that described the spatial aspects of the land base (e.g., elevation, forest cover, management units, roads), as well as tables (e.g., volume yield tables, harvest flow) and parameters (e.g., harvest level) containing information that is not tied to a specific piece of land. Time is modeled in 1-year or 10-year steps, with a time horizon of 250 to 400 years.

Outputs included text files that recorded various aspects of the condition of the land base (e.g., the growing stock or age-class distribution) and spatial time series (e.g., stand ages). The MLM simulates specified processes by projecting initial landscape conditions forward through time using stochastic techniques (i.e., by using the probabilities of change from one state to another). However, it does not determine optimal solutions. Thus, each model run may produce different results and the model must be run several times to determine averages and ranges for each scenario being modeled. The model's users must then compare these results with those of other simulations based on different parameters. The overall model design is shown in Figure 3.3, in which landscape states are shown in the middle, process submodels are shown as ovals, and output files are shown as gray cylinders. An arrow emanating from a

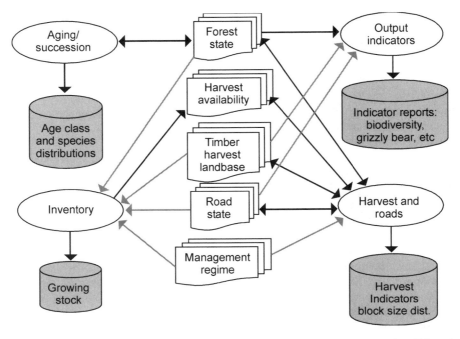

Figure 3.3. The overall conceptual design of the Morice Landscape Model (drawn based on Fall et al. 2004a). Each main modeled process is shown as an oval; the main components of the landscape state (represented as spatial data layers and nonspatial tables) are shown in the center and output files are shown as gray cylinders. Arrows indicate that a process depends on or modifies the connected landscape state.

process indicates that the process creates an output or modifies a state variable, whereas an arrow emanating from a state category indicates that the state influences the behavior of a process.

3.4.2. Process Models

The MLM simulation of landscape change included forest growth, stand-replacing disturbance, forest harvesting, and road construction (Figure 3.3). Within-stand disturbances caused by diseases, insects, and windthrow were not explicitly modeled; however, their timber-related impacts are accounted for in estimates of the volume harvested.

Forest growth. The forest growth submodel increments the stand age in forested cells to a maximum age (650 years) and updates any changes in species due to planting. Succession was modeled based on diagrams of vegetation pathways that captured the trends in species shifts over time on different sites after different events; these were developed by means of expert consultation and workshops (Beukema and Pinkham 2001). The stand volume at a given age for a given site type was estimated

A Collaborative, Iterative Approach

by looking up the corresponding value in a yield table, and was summarized to obtain indicators of growing stock.

Natural disturbance model. Stand-replacing natural disturbance was modeled using disturbance rates and patch-size distributions for the area's biogeoclimatic zones based on an analysis of historic disturbance levels for the area (Steventon 2002). This top-down, empirical approach captures all agents of stand-replacing natural disturbance, and was used to help ascertain appropriate targets for ecosystem-based management objectives. The disturbance agents were primarily fire, mountain pine beetle (*Dendroctonus ponderosae* Hopkins), spruce beetle (*Dendroctonus rufipennis* Kirby), and western balsam bark beetle (*Dryocoetes confusus* Swain).

Harvesting model. The harvesting submodel was adapted from a prior spatial timber-supply model constructed in SELES that captured the same management regimes, assumptions, and data requirements as the aspatial timber supply model Forest Service Simulator (FSSIM) used for timber supply analysis by the British Columbia Ministry of Forests (BCMOF 2002). The submodel was extended to include spatial constraints such as spatial blocks (i.e., barriers to a process), road access, and block adjacency. A scenario to capture current forest management policy was calibrated against an aspatial analysis done as part of the province's timber supply review process using FSSIM (BCMOF 2002). This calibration step was key to the acceptance of MLM by foresters who were unsure of the model's ability to perform a realistic timber supply analysis.

In general, harvest blocks were limited to eligible land, such as accessible stands older than the minimum harvest age. The start points for cut blocks were selected from among the eligible sites based on stand age (with preference increasing as stands aged beyond the harvestable age); the blocks then grew to encompass neighboring cells until a preselected block size was reached. Harvest effects include extraction of merchantable volume, resetting of stand age, constructing roads (for sites accessible from the ground), and updating of tracking variables (e.g., the annual area harvested). Blocks are simulated sequentially within a period until the harvest target for the period is met.

Road access. The logging submodel explicitly connects cut blocks to the main road network by identifying straight-line spur roads to the nearest mapped road or previous spur road. If a proposed mapped road segment is connected to a spur, it is activated. This method of modeling road development accounts for feedback between current limitations on access and the road building required to permit harvesting of certain areas, and thereby reduces access limitations over time.

3.4.3. Indicator Models

Indicator models were designed by the researchers to summarize and produce customized information used by the resource experts to assess timber and ecological values. As such, they were developed in close collaboration with the resource experts, so that each expert shared ownership of that portion of the MLM. This improved buy-in from the wide variety of experts, and increased their confidence in subsequent analyses and presentations to the Land and Resource Management Plan planning group.

3.4.4. Outcomes

The Morice Land and Resource Management Plan planning group reached consensus early in 2004, and the final agreement was passed to the provincial government's cabinet ministers, who are the top-level stewards of public land in British Columbia. Further application of the MLM was used to help support a socioeconomic impact assessment by government experts, who in turn advised the politicians. In the spring of 2004, the plan was accepted and the implementation stage began. Implementation involved additional negotiations with First Nations, changes in legislation, the initiation of management plans for new zones, operational restructuring for affected people and industries, and the design of monitoring systems to assess the plan's objectives as implementation unfolded. The reliance on the MLM in the final political stage of the process demonstrates how this approach facilitated the transfer of analytical results from an exploratory stage by the planning group (which focused on the relative costs and benefits of alternative plans) to final analysis (which focused on the absolute costs and benefits of the final plan).

3.4.5. Obstacles and Lessons Learned

The importance of transparency and adequate communication in this process cannot be overstated. It was essential to walk interested members of the government technical team through the main model assumptions. This was a challenging step, but was necessary to ensure that the planning group would have confidence in the results presented by the resource experts. A second challenge related to managing expectations and maximizing shared learning of capabilities of the system and the model. On the one hand, the researchers had to strive to provide the flexible, customized information required to support the planning process, and on the other hand it was critical to communicate clearly about items that were not feasible to include (e.g., due to a lack of time or data). A third challenge related to timing: preliminary analysis of the preplan management and the final plan analysis had to be quite detailed, but neither was constrained by time. However, analysis of the trade-offs in intermediate versions of the plan, especially toward the end of the planning negotiations, had to be done very quickly (i.e., within days, and sometimes hours) to provide information in a relevant time frame. When deadlines could not be met, the information could not be used.

3.5. CASE STUDY 2: MANAGING FIRE RISK IN THE WILDLAND–URBAN INTERFACE

Mitigating the risk of wildfire has become an urgent issue for the managers of many North American forests (Finney and Cohen 2002). Wildfire risk is a consequence of complex interactions among natural disturbance regimes, vegetation management activities (e.g., timber management, fuel reduction), and the increased presence of people in forested landscapes. Portions of the Lakewood Unit of the Chequamegon-Nicolet National Forest (Wisconsin, USA) feature both highly flammable jack pine (*Pinus banksiana* Lamb.) and red pine (*Pinus resinosa* Ait.) and a relatively high proportion of privately owned land within the forest that have experienced rapid exurban development in recent decades, a trend that is expected to continue. Recent research by Sturtevant and Cleland (2003) indicates that the probability of forest fire ignitions in Wisconsin is primarily a function of housing density, whereas the probability of large fires depends more on the ecosystem properties that control fire spread, such as soil water retention and the flammability of the vegetation. Fire management officers and land planners were seeking long-term guidance on forest management strategies to mitigate the risk of fire damage to timber and private property in the Lakewood Unit, a fire-prone, mixed-ownership region of the National Forest. The critical issue was the rapid increase in exurban development intermixed with fire-prone federal lands.

Two of us (Sturtevant and Gustafson) began collaborating with fire-management officers responsible for the Chequamegon-Nicolet National Forest. Our purpose was to describe the available tools for landscape simulation and identify critical issues that might be addressed with such tools. We began the collaborative, iterative process with a simple discussion between ourselves and decisionmakers for the National Forest. The managers explained the management situation described earlier in this section, and expressed their need for a scientific basis on which to generate and evaluate specific management actions they could take to reduce the risk of wildfire and protect the human communities within the wildland–urban interface. We described several modeling tools that could meet such a need, and outlined the pros and cons of each. After some discussion, the team decided that a combination of tools would be required. LANDIS would estimate fire risk by accounting for the interactions among vegetation-management treatments, forest succession, natural disturbance, and human-caused ignitions. However, LANDIS cannot currently predict the expansion of human populations through time. This ability was deemed critical to meet the future decision-support needs of the managers, but was deferred to a later phase of the project. In the meantime, we used the current extent of exurban development.

We spent the following months preparing for the first workshop, in which we would begin the collaborative technology transfer. The original participants in the discussion collectively identified additional key participants, including land managers and planners from the National Forest (the Fire Management Officer, District Ranger, and District Project Coordinator), resource experts (two silviculturists and the District Fire Management Officer), and researchers familiar with the modeling tools (two LANDIS developers, two LANDIS users, and a human demographer). Finally, we acquired the input data needed to conduct prototype LANDIS simulations, including data on current forest conditions, current harvest practices, the region's fire regime, and the current pattern of exurban development.

There were four main objectives of the first workshop. First, we needed to clearly articulate the objective of the collaboration based on input from all participants. In this case, our objective was to investigate the fire risk associated with various forest-management alternatives and their interaction with exurban development within the Lakewood Unit. Second, the researchers explained the technology (i.e., LANDIS) that would be applied to meet the stated objective. Third, the researchers presented prototype LANDIS simulations, and used the results to spark discussions with the resource experts on several areas of uncertainty in the prototype. For example, we discussed the relevant and important tree species in this ecosystem, the role of wetlands in landscape-scale fire behavior, the critical drivers responsible for fire spread, local fire-suppression tactics (e.g., the use of roads as fire breaks), and the forest composition on private land (i.e., where data were lacking). Finally, the land planners provided guidance on how to implement the current forest management plan within LANDIS, and discussed potential strategies for mitigating fire risk.

The research team spent several months incorporating the recommendations and data resources provided by the broader group into a LANDIS simulation run. During this time, we consulted regularly with the local resource experts for clarification and refinement of the simulation parameters. Our efforts at this stage were devoted to developing realistic LANDIS simulations. Simulation runs projected the forest's composition, spatial pattern of fuels, and fire risk in the Lakewood Unit over the next 250 years based on the current forest management plan and on the assumption of no additional exurban development.

The next iteration of our collaboration was conducted in a second workshop with the entire group and the results of the initial simulations were presented. The resource experts and planners then provided feedback on the realism and utility of the model results. Our demonstration was followed by a brainstorming session to develop several novel, spatially explicit fire-mitigation scenarios. The alternatives were constrained by the current Land and Resource Management Plan, which established the broad management directions for the National Forest but allowed considerable latitude in the implementation tactics. For example, strategic conversion of existing coniferous stands and establishment of new coniferous stands were both designed to minimize the adjacency between human ignition sources and coniferous fuels and to reduce the overall risk of wildfire spread across the landscape. Because the input maps and parameter files were already in place, we could quickly implement the scenarios for mitigating fire risk.

The effectiveness of these strategies was evaluated during the third iteration of the process to gain insight into the effects of specific elements of the risk-mitigation strategies. This allowed the managers to identify the most effective strategy for further evaluation and refinement.

3.5.1. Outcomes

One important outcome of the process was that each group in the triad (Figure 3.1) gained valuable insights from the other two groups, and this helped them to contribute more usefully to the final decision. Another valuable outcome was that the modeling technology was used to support a critical management decision.

But the most important outcome is that the team will continue to develop a sound fire and fuel mitigation strategy for the Lakewood Unit. The temporal and spatial distribution of fire risk predicted by the LANDIS model helped the team to develop novel ideas for mitigating fire risk that might not have been apparent without the model projections for reference. The LANDIS technology thus allowed the team to evaluate the effectiveness of alternative strategies both spatially and quantitatively. The strategy that is eventually adopted is expected to be superior to one that would have been developed without the decision support made possible by the collaborative, iterative approach to transferring LANDIS technology.

3.5.2. Obstacles and Lessons Learned

Our collaborative approach to technology transfer required effective communication among the three knowledge groups (Figure 3.1), using the landscape model as a means of focusing critical discussions and a series of workshops as the primary means of information exchange. Well-organized workshops that foster communication were therefore a key to our success. A potential obstacle in reaching that goal lies in finding the appropriate balance between general discussions and examinations of the specific details of the model. Discussion topics should be organized around key modeling assumptions or uncertainties, but should remain sufficiently broad to encourage real feedback from all three knowledge groups. In turn, these general discussions must be followed by more specific question-and-answer periods that will be used to actually parameterize the model. Nonetheless, the tendency of scientists organizing a technology-transfer workshop is to dwell exclusively on the model details (i.e., the technology), an approach that can effectively discourage more fundamental input from the other two knowledge groups. In our experience, the encouragement of more general discussions allows participants to identify areas of uncertainty that are apparent only to those with more direct knowledge of the system to be modeled.

A second obstacle that we encountered was the difficulty in clearly communicating among the different knowledge groups, each of which had its own terminology and frame of reference. For example, the fuel module of LANDIS 4.0 was designed to incorporate local expert opinion so as to guide the interaction between fire and fuel patterns (He et al. 2004). However, because local fire experts did not participate in the development of the LANDIS fuel classes, we found it difficult to translate their expert experience into actual parameters. In the end, we provided output comparisons between LANDIS and the modeling tools the fire experts were already familiar with (i.e., BehavePlus, Andrews et al. 2003; FARSITE, Finney 1998), thereby effectively calibrating the LANDIS model to fit their experience. Though this effort took additional time, it both informed the researchers and improved the confidence of local experts in the output of the LANDIS model. Another method for ensuring effective communication among groups is to provide a brief synopsis following each workshop, and circulate it among the group participants for comment. This provides an opportunity to correct any misinterpretations of workshop discussions before they are incorporated in the model.

Because our simulations assumed a static human presence in the study landscape, the effects of future development were not incorporated into the predictions. The research team brainstormed solutions to the human component of the question, and identified an appropriate simulation tool (the "planning support system") to fill this need. Planning support systems are integrated packages of analytical tools based on geographical information systems technology that perform three critical tasks: conducting an analysis of land suitability, projecting future land-use demand, and allocating the projected demand to suitable locations. Although these systems do not predict future conditions exactly, they do attempt to determine future conditions given certain policy choices and development assumptions. We have decided to use such a tool to provide the human development projections that will be incorporated in future LANDIS simulations.

3.6. COMPARISON OF THE CASE STUDIES

One of the clear differences between the two case studies lies in the development of a new model versus the use of an existing model. Technology transfer performed during the collaborative development of a new model offers two main advantages. First, all participants share intellectual ownership of the technology, and this increases their confidence in the model results. Second, developing a new model ensures that the tool fits both the available data and the specific questions that must be answered by the management application (Fall et al. 2001). In the second case study, much of the development time was devoted to transforming the available data into the inputs required by LANDIS. The choice between creating a new model and using an existing one thus depends on how well available models can address the question at hand, and on the time required to implement an existing model compared with creating a new one. Software tools such as SELES (for landscape models) and STELLA (Costanza et al. 1998; for stock and flow models, which represent a system as compartments or stocks of entities and fluxes or flows of matter or energy between stocks) are decreasing the time required to create new models, and are therefore conducive to the collaborative approach to technology transfer that is described in this chapter. Previously published models have the advantage of the additional scientific rigor that results from peer review. In the second case study, less energy was required to validate the successional dynamics of LANDIS because the model had been tested extensively in various temperate ecosystems around the world (Mladenoff 2004).

Although these case studies differed markedly in terms of the objectives of the decisions being supported, the ecology and disturbance regimes of the landscape under study, and the technology used for the decision-support model, a number of common themes relate directly to the iterative, collaborative process that we have proposed. One critical component of both case studies was the active participation of all knowledge groups in developing the landscape model (Case Study 1), parameterizing the landscape model (Case Study 2), and developing and evaluating relevant scenarios (both case studies). Such active participation inspired confidence in the technology by transferring intellectual ownership from the researcher to the other two groups. Workshops and face-to-face discussions on various aspects of the problem (decision objectives, local knowledge, capabilities and limitations of the technology) facilitate participation. Note that it is not necessary that all aspects of the model be transparent to all participants in the process. Different resource experts can verify the assumptions for different aspects of the model, and can transfer their confidence in those assumptions to the rest of the participants. Such synergistic participation minimizes the time investment required from each member of the collective group. However, it does increase the time investment by the researcher in the technology-transfer process. Justification for this additional time investment comes from the increased future independence of the managers when they use the technology, and the increased knowledge gained from the other two groups that researchers can apply to future technology transfer.

In both case studies, once the analytic technology had been selected, discussions focused primarily on the conceptual model of the system and the information required as inputs for this model, because this is the level at which meaningful discussion among the three parties is possible. Too much focus on the tool draws attention away from the main objective, and may limit accessibility to the process for people who lack sufficient technical training. Both case studies demonstrated flexibility in how the collaborative, iterative approach was applied—it was not a recipe to be rigidly followed. Rather, the appropriate timing of workshops and levels of involvement by the various parties must emerge during the course of a project, thereby enhancing mutual learning and ensuring adequate levels of communication.

There is a risk that researchers may lose their objectivity by working so closely with members of the planning team. Also, the distinction between the local resource experts and the planning team may become blurred when members serve in both capacities. Care must be taken to avoid tweaking the model inputs and assumptions to fit some preconceived notion of the results. To ensure that the technology is not abused, the modeler must clearly document the model's assumptions and limitations, and the appropriate methods of interpreting the outputs. During technology transfer, these potential pitfalls imply that any project that applies complex systems models requires at least one person to fill the researcher's role and ensure that all participants are aware of the pitfalls and can respond appropriately.

3.7. CONCLUSIONS

Based on the experiences described in this chapter, we conclude that the collaborative, iterative approach to technology transfer provides several important benefits that are lacking in traditional buyer–seller approaches. First, the collaborative component encourages the establishment of long-term working relationships. This is important because it develops a mutual commitment to a successful outcome by fostering a shared vision and shared goals. It also allows a shift from a focus on decisions as discrete events to a focus on decisions as a continuous process that fits within the context of adaptive management. The focus of all parties is on the outcomes rather than on the process of technology transfer. In contrast, the traditional buyer–seller relationship between researchers and managers has evolved into a marketing game in which the sale is more important to the researcher than the effectiveness of the management decisions (i.e., the pressure to report technology transfer may encourage researchers to focus on their self-interest). A collaborative approach serves to correct these perverse incentives and break down some of the walls that have separated managers and researchers by increasing the likelihood of a mutual success.

Second, the iterative component of the approach serves to progressively improve the quality and relevance of the model as a decision-support tool. Model parameters and inputs are improved by the reality checks provided by managers and resource experts during each iteration. Because the model and its results are described and discussed at some length as an integral part of the iterative process, the managers become increasingly educated about the technology, and the model becomes much less likely to be perceived as a mysterious black box. This inspires more confidence in and understanding of the results, thereby increasing their value for decision support. The extended interactions that result from this approach help all parties to learn and develop a more mature understanding of the overall picture. Each party is in effect providing on-the-job training for the others, and this training should enhance the effectiveness and efficiency of future partnerships with the same team or with new teams established for other purposes.

Third, in this approach, all parties have a vested interest in the success of the other parties. The approach is framed in terms of the outcome rather than in terms of technology transfer per se. When the outcome is achieved, all parties can claim success. Managers can then take advantage of the latest modeling technology to obtain sound and relevant support for their decisions. Their decisions will become more defensible, and they will be more likely to achieve their management objectives. Resource experts will provide critical input to the process, and will therefore have played a key role in shaping the management decision. Researchers will have successfully transferred their technology to a management environment, and know that their technology will make a difference "on the ground."

The collaborative, iterative approach to technology transfer can be applied to any complex technology used to support natural-resource management decisions. The main feature of the approach is a sustained partnership that changes the technology-transfer paradigm from a buyer–seller mentality to an outcome-based model in which all parties can win. The approach is structured so that all parties achieve success when the outcome is achieved, and this provides adequate incentives to encourage meaningful collaboration. The approach itself is not complex, but it facilitates the transfer of complex technology by keeping those who understand those complexities attached to the technology. We believe that in many cases, this is the only way to achieve successful application of such technology to "on the ground" management decisions.

Acknowledgments

We thank Deahn Donnerwright, Don Riemenschneider, Linda Parker, Jim Sanders, Duane Lula, Volker Radeloff, Jiquan Chen, and the anonymous reviewers for thoughtful reviews that helped us improve the manuscript.

LITERATURE CITED

Ahern J (2002) Spatial concepts, planning strategies and future scenarios: a framework method for integrating landscape ecology and landscape planning. In: Klopatek J, Gardner R (eds) Landscape ecological analysis: issues and applications. Springer, New York, pp 175–201.

Allee V (1997) The knowledge evolution: expanding organizational intelligence. Butterworth-Heinemann, Boston.

Alverson WS, Waller DM, Solheim SL (1988) Forests too deer: edge effects in northern Wisconsin. Conserv Biol 2:348–358.

Andrews PL, Bevins CD, Seli RC (2003) BehavePlus fire modeling system, version 2.0: User's guide. USDA For Serv, Rocky Mountain Res Stn, Ogden, Utah. Gen Tech Rep RMRS-GTR-106WWW.

Baker WL (1989) A review of models of landscape change. Landscape Ecol 2:111–133.

BCMOF (2002) Morice timber supply area analysis report. British Columbia Ministry of Forests, Victoria, BC.

Beukema SJ, Pinkham CB (2001) Vegetation pathway diagrams for the Morice and lakes innovative forest practices agreement. ESSA Technologies Ltd., Vancouver, BC. Project Rep IFPA No. 451.01.

Boutin S, Hebert D (2002) Landscape ecology and forest management: developing an effective partnership. Ecol Appl 12:390–397.

Costanza R, Duplisea D, Kautsky U (1998) Ecological modelling and economic systems with STELLA. Ecol Model 110:1–4.

Fall A, Fall J (2001) A domain-specific language for models of landscape dynamics. Ecol Model 141(1–3):1–18.

Fall A, Daust D, Morgan D (2001) A framework and software tool to support collaborative landscape analysis: fitting square pegs into square holes. Trans GIS 5(1):67–86.

Fall A, Morgan D, Edie A (2004a) Morice landscape model. Report to the Morice Land and Resource Management Planning Process, BC Min Sustainable Resour Manage, Victoria, BC.

Fall A, Shore TL, Safranyik L, Riel WG, Sachs D (2004b) Integrating landscape-scale mountain pine beetle projection and spatial harvesting models to assess management strategies. Proc Mountain pine beetle symposium: challenges and solutions. Shore TL, Brooks JE, Stone JE (eds), 30–31 October 2004, Kelowna, British Columbia. Nat Resour Canada, Can For Serv, Pacific For Centre, Victoria. Inf Rep BC-X-399.

Finney MA (1998) FARSITE: fire area simulator—model development and evaluation. USDA For Serv, Rocky Mountain For Range Exp Stn, Fort Collins. Res Pap RM-4.

Finney MA, Cohen JD (2002) Expectation and evaluation of fuel management objectives. USDA For Serv, Rocky Mountain For Range Exp Stn, Fort Collins. USDA For Serv Proc RMRS-P-29, pp 353–366.

Fleming RA, Candau JN, McAlpine RS (2002) Landscape-scale analysis of interactions between insect defoliation and forest fire in central Canada. Climatic Change 55:251–272.

Gustafson EJ, Shifley SR, Mladenoff DJ, He HS, Nimerfro KK (2000) Spatial simulation of forest succession and timber harvesting using LANDIS. Can J For Res 30:32–43.

Gustafson EJ, Zollner PA, Sturtevant BR, He HS, Mladenoff DJ (2004) Influence of forest management alternatives and landtype on susceptibility to fire in northern Wisconsin, USA. Landscape Ecol 19:327–341.

Hargrove WW, Gardner RH, Turner MG, Romme WH, Despain DG (2000) Simulating fire patterns in heterogeneous landscapes. Ecol Model 135:243–263.

He HS, Mladenoff DJ (1999) Spatially explicit and stochastic simulation of forest-landscape fire disturbance and succession. Ecology 80:81–99.

He HS, Mladenoff DJ, Boeder, J (1999a) An object-oriented forest landscape model and its representation of tree species. Ecol Model 119:1–19.

He HS, Mladenoff DJ, Crow TR (1999b) Linking an ecosystem model and a landscape model to study forest species response to climate warming. Ecol Model 114:213–233.

He HS, Shang BZ, Crow TR, Gustafson EJ, Shifley SJ (2004) Tracking forest fuel dynamics across landscapes—LANDIS fuel module design. Ecol Model 180:135–151.

Liu J, Ashton PS (2004) Simulating effects of landscape context and timber harvest on tree species diversity. Ecol Appl 9:186–201.

Manseau M, Fall A, O'Brien D, Fortin M-J (2002) National parks and the protection of woodland caribou: a multi-scale landscape analysis method. Res Links 10(2):24–28.

Mladenoff DJ (2004) LANDIS and forest landscape models. Ecol Model 180:7–19.

Mladenoff DJ, He HS (1999) Design, behavior and application of LANDIS, an object-oriented model of forest landscape disturbance and succession. In: Mladenoff DJ, Baker WL (eds) Spatial modeling of forest landscape change: approaches and applications. Cambridge Univ Press, Cambridge, UK, pp 125–162.

Morgan D, Daust D, Fall A (2002) North Coast landscape model. Report to the North Coast Land and Resource Management Planning Process, B.C. Min Sustainable Resour Manage, Victoria, BC.

Sklar FH, Costanza R (1991) The development of dynamic spatial models for landscape ecology: a review and prognosis. In: Turner MG, Gardner RH (eds) Quantitative methods in landscape ecology. Springer-Verlag, New York, pp 239–288.

Steventon D (2002) Historic disturbance regimes of the Morice and Lakes timber supply areas. BC Min For, Smithers. Unpubl Discussion Rep.

Sturtevant BR, Cleland DT (2003) Human influence on fire disturbance in northern Wisconsin. Proc 2nd Int Wildland Fire Ecol and Fire Manage Congr, 18–20 November 2003, Orlando, Florida.

Sturtevant BR, Gustafson EJ, Li VW, He HS (2004a) Modeling biological disturbances in LANDIS: a module description and demonstration using spruce budworm. Ecol Model 180:153–174.

Sturtevant BR, Zollner PA, Gustafson EJ, Cleland D (2004b) Human influence on fuel connectivity and the risk of catastrophic fire in mixed forests of northern Wisconsin. Landscape Ecol 19:235–253.

Teich GMR, Vaughn J, Cortner HJ (2004) National trends in the use of forest service administrative appeals. J For 102:14–19.

Turner MG (1989) Landscape ecology: the effect of pattern on process. Annu Rev Ecol Syst 20:171–197.

Turner MG, Wu Y, Wallace LL, Romme WH, Brenkert A (1994) Simulating winter interactions among ungulates, vegetation, and fire in northern Yellowstone Park. Ecol Appl 4:472–496.

Weathers KC, Cadenasso ML, Pickett, STA (2001) Forest edges as nutrient and pollutant concentrators: potential synergisms between fragmentation, forest canopies, and the atmosphere. Conserv Biol 15:1506–1514.

With KA (2002) The landscape ecology of invasive spread. Conserv Biol 16:1192–1203.

Yang J, He H, Gustafson EJ (2004) A hierarchical statistical approach to simulate the temporal patterns of forest fire disturbance. Ecol Model 180:119–133.

4

Development and Transfer of Spatial Tools Based on Landscape Ecological Principles: Supporting Public Participation in Forest Restoration Planning in the Southwestern United States

Haydee M. Hampton, Ethan N. Aumack, John W. Prather*, Brett G. Dickson, Yaguang Xu, and Thomas D. Sisk

4.1. Introduction
4.2. ForestERA Background and Objectives
4.3. Transfer of Forest Landscape Ecological Knowledge
    4.3.1. Educating Stakeholders on the Application and Utility of Analyses Based on Forest Landscape Ecology
    4.3.2. The ForestERA Spatial Decision-Support System
    4.3.3. The Value and Function of Collaboration during Decisionmaking
4.4. Knowledge Transfer Mechanisms
    4.4.1. Engaging Stakeholders in Project Development
    4.4.2. Delivering Data and Tools Designed to Meet Stakeholder Needs

---

*John Prather passed away February 20, 2006. We dedicate this chapter to him.

HAYDEE M. HAMPTON, JOHN W. PRATHER, YAGUANG XU and THOMAS D. SISK • Northern Arizona University, Center for Environmental Sciences and Education, NAU Box 5694, Flagstaff, AZ 86011, USA. ETHAN N. AUMACK • Grand Canyon Trust, 2601 N. Fort Valley Road, Flagstaff, AZ 86001, USA. BRETT G. DICKSON • Colorado State University, Department of Fishery & Wildlife Biology, Fort Collins, CO 80521, USA.

4.4.3. Letting the Stakeholders Drive: Testing Forest Landscape Ecological Tools during Their Development
4.4.4. Transfer of Collaborative Planning Processes
4.4.5. Supporting the Learning Process Using ForestERA Products
4.5. Successes and Challenges in Knowledge Transfer
    4.5.1. Successful Use of the ForestERA Data and Tools
    4.5.2. Overcoming Obstacles
    4.5.3. Changes for Future Efforts
    4.5.4. What Others Can Learn from Our Experience
4.6. Concluding Remarks
    Literature Cited

4.1. INTRODUCTION

More than half of the North American continent was covered with forests when European explorers and colonists first established permanent settlements, and for centuries the forests seemed inexhaustible. In this context of seemingly limitless land, abundant timber resources, and ongoing westward expansion, the United States began a farsighted experiment—holding vast tracts of undeveloped land as public property and managing them under federal and state authority for the benefit of all citizens (for details, see Wilkinson and Anderson 1987). This experiment, bold in scope and visionary in concept, has retained forest cover over most of these lands, despite overzealous exploitation that liquidated virtually all southwestern old-growth forests and that degraded forest conditions in much of the region (Behan 2001). Yet on the whole, the public lands experiment in the United States has been a conservation success. Most of the National Forest System lands remain in a seminatural state, in contrast to productive private forests, which have largely been converted to agroforestry. Moreover, citizens retain a powerful voice in how public forest lands are managed.

Sustaining public engagement in the twenty-first century, with an ever-more politicized planning and management process, has placed increasing pressure on efforts to engender meaningful public engagement and pursue science-based management. In some cases, public involvement has been tokenized, with perfunctory meetings and comment periods, whereas the battle between local and national interests has become heated, nasty, and sometimes even violent (e.g., Durbin 1999). Yet the long-term success of the public forest experiment depends on rekindling and sustaining the public engagement that has guided forest management through intense controversies involving clearcutting, road-building, conservation of endangered species, and privatization, to name only a few points of recent dispute. Though messy, this public engagement in forest management, ranging in form from written comments to high-profile litigation, provides critical input into the workings of federal and state bureaucracies charged with managing forests in the public interest.

Supporting Public Participation

This input is particularly important when scientific uncertainty and conflicting values cloud decisionmaking. In an era of rapidly growing human populations and increasing demands on forest ecosystems, the resolution of emerging conflicts demands increased public access to the best science, and a process for engaging a broad range of citizens in science-based discourse (Sarewitz 2004). This challenge—to make science accessible and practical—has been undertaken in different forms across the continent. In this chapter, we offer a case study from the semiarid Southwest, where the threat of wildfire has focused public attention and engendered bitter debate about the future of public forests and how they will be managed.

After providing background information on our case study in Section 4.2, we present the *what* and the *how* of knowledge transfer in forest landscape ecology as it applies to our project: what was transferred, including data, models, analytical and statistical tools, and collaborative processes (Section 4.3). In Section 4.4, we examine how the project was designed for effective transfer via a multiyear dialog that developed trust among collaborators and that allowed meaningful engagement when the scientific tools were ready to support a more focused public process. We also review the implementation of specific transfer mechanisms through several planning processes, including the knowledge transfer related to the Western Mogollon Plateau Adaptive Landscape Assessment, which brought together more than 100 forest managers, scientists, public officials, and engaged citizens in a series of workshops to carry out the assessment. Finally, we summarize the lessons learned (Section 4.5) to help others who may wish to undertake similar endeavors.

4.2. FOREST*ERA* BACKGROUND AND OBJECTIVES

Ponderosa pine (*Pinus ponderosa* Laws.) is widely distributed at higher elevations in the southwestern United States. In Arizona and New Mexico, it is the dominant species across an 8.4-million-acre (3.4 million ha) arc stretching from the Gila National Forest in New Mexico, across Arizona's Mogollon Plateau, and up to the Kaibab Plateau (the gray region in Figure 4.1). Across this vast forest, as across most of the intermountain western United States, heavy logging, livestock grazing, and fire suppression have converted the forest structure from relatively open conditions characterized by fewer, larger, fire-resistant trees to denser stands of smaller trees (Allen et al. 2002, Covington and Moore 1994, Dahms and Geils 1997). Concomitant with this shift in structure, and partially responsible for it, was a radical change in fire frequencies. Prior to European settlement of the Southwest, fire was a frequent ecological process throughout the ponderosa pine forest type, as it typically burned across the ground, consuming duff, dead and downed wood, and tree seedlings, but only infrequently spread through the forest canopy. Fire burned these forests regularly, with estimates of return intervals ranging from 3 to 12 years in most studies (Allen et al. 2002; Covington et al. 1997; Swetnam and Betancourt 1998). In the early twentieth century, as the economic value of old-growth forests was being realized by a booming timber industry, fire suppression became a primary

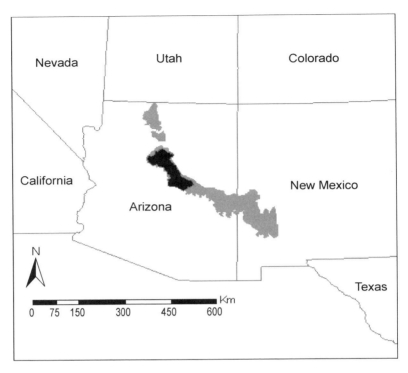

Figure 4.1. The Western Mogollon Plateau study area (black) is located in northern Arizona, within the world's largest continuous expanse of ponderosa pine forest (gray).

objective of forest managers, and the capacity to contain and extinguish fires grew rapidly. By the 1920s, fires in the ponderosa pine forest type had been drastically curtailed, and seedling recruitment exploded, leading to the present heavily stocked forests. As trees matured and suppression activities prevented ground fires from consuming the accumulating fuels, the nature of fire in the ponderosa pine ecosystem changed dramatically. Fires burned hotter, spread faster, and killed more mature trees than the frequent ground fires of earlier times. More recently, there has been a marked increase in the size and severity of forest fires in the Southwest, whether ignited by human or natural causes (Dickson et al. 2006) This shift in fire behavior coincided with increasing human populations and increasing development in the forest, resulting in increased risk to life and property and elevating the management of public forest lands to the national political agenda.

Currently, considerable policy effort and public expenditure focuses on revising the management of southwestern ponderosa pine forests. Abatement of fire risk, protection of communities, watershed protection, conservation, timber harvesting, grazing, and recreation present a broad and confusing array of objectives that are interwoven with public opinion, liability issues, and competition for political power.

In this policy environment, engaging a vigorous, constructive, and informed public discourse becomes challenging. Furthermore, despite a rapidly increasing scientific understanding of forest ecology, science is often sidelined during heated policy debates, and technical arguments resonate less and less with the citizens, technocrats, and decisionmakers who define the terms of the debate. Yet history suggests that informed and inclusive public debate is vital because it clarifies disputed values, provides a critique of current management paradigms, and generates new ideas that offer alternative solutions for seemingly intractable management problems.

The intense fire season of 2000 burned more than 600 000 acres (243 000 ha) in Arizona and New Mexico, including part of the city of Los Alamos, forcing an evacuation that raised forest management questions that reverberated in Washington, D.C. The ensuing debate resulted in emergency appropriations in fiscal year 2001 by the U.S. Congress (Department of Interior and Related Agencies Conference Report 106-914) to fund progressive approaches to the emerging wildfire crisis that had been brought on by mismanagement and drought in the Southwest. The emergency appropriations bill called for "an adaptive ecosystem analysis of ponderosa pine and related forests as a prototype for larger ecosystem analyses, and to fill the gap between project or district/forest level analyses and regional analyses to support future operational scale treatments." This mandate reflected the growing realization that sound planning should occur at the same scale as the defining disturbances—such as wildfire—an idea that emerged from advances in theoretical ecology 15 years earlier (Pickett and White 1985). Funding for this project was awarded to the Ecological Restoration Institute of Northern Arizona University, located in Flagstaff (Arizona), in the heart of the ponderosa pine ecosystem. In 2002, we initiated the ForestERA project with funding provided to the Ecological Restoration Institute as a subgrantee under the appropriations bill. Our study region comprised the entire 8.4 million acres (3.4 million ha) of ponderosa pine forest (Figure 4.1). We first applied in-depth analysis to a 2.1-million-acre (850 000 ha) region of relatively homogeneous forest surrounding Flagstaff, which we refer to hereafter as the Western Mogollon Plateau.

Over a 2-year period, our team developed new spatial data on forest conditions, augmented these with existing data sets that met our quality standards, and developed landscape-scale ecological models to predict how the forest's condition would influence variables of interest to planners and policymakers, such as fire risk (the likelihood of occurrence), fire hazard (the expected magnitude of the damage should a fire occur), habitat quality for a range of wildlife species, and watershed conditions. These analytical tools allow the exploration of various management scenarios in a spatially explicit digital environment, and provide a means for assessing and comparing the predicted consequences of these alternative scenarios for forest values of broad public interest. This scientific content represents the traditional component of the knowledge whose transfer is the principal subject of this chapter. However, in tandem with the ecological science and technology, we invested heavily in the development, trial, refinement, and implementation of a public process wherein the science of Northern Arizona University's Forest Ecosystem Restoration

Analysis (ForestERA) project was delivered to the broadest possible audience of potential users in a manner that was accessible, relatively user-friendly, and practical. Our objective was to provide a robust scientific and technical capacity for a diverse audience, comprising the full range of citizens, scientists, and public servants who were interested and engaged in issues of forest conservation and management. It is the transfer of this integrated capacity—the data, models, scenario-assessment tools, and public process—that is informed by, but not controlled by, the rigorous science. Our assumption was that, when successful, these efforts would help identify forest management scenarios that responded to differing public values and competing interests and led to on-the-ground management actions that would restore appropriate ecosystem structure and function to degraded ponderosa pine forests.

Throughout the ForestERA project, we have labored to sustain an ongoing dialog with a range of stakeholders in the management of ponderosa pine forests in Arizona and New Mexico. This communication helped focus the objectives for our project, which included efforts to:

- Develop and distribute spatial data layers and the tools needed to access, explore, analyze, and display them. These spatial layers present the locations and values of specific parameters, such as the presence or absence of a particular wildlife habitat, in geographical information system (GIS) software. This work lies at the core of the ForestERA project, and occupied the majority of our time and resources.
- Assist stakeholders in incorporating a landscape perspective in their consideration of ongoing and future forest planning. These efforts included the work of state and federal agencies in planning and implementing forest thinning, prescribed burning, and fire suppression, as well as decisionmaking regarding wildlife management, watershed protection, and community protection.
- Help stakeholders engage in a collaborative planning process that will allow them to explore and understand each other's point of view, compare alternative approaches for meeting both shared and conflicting goals, and forge a perspective for future management that, even when it did not resolve all differences, at least moved toward common ground.

Meeting each of these overarching objectives required an integrated approach to landscape analysis that fused public process with spatial modeling.

4.3. TRANSFER OF FOREST LANDSCAPE ECOLOGICAL KNOWLEDGE

The knowledge related to forest landscape ecology that we are transferring to stakeholders in the ForestERA project falls into three general categories: education, spatial data and tools, and collaborative processes for fostering public participation in forest planning. An essential and ongoing educational component communicates the utility and application of landscape-scale spatial data and analyses. To increase stakeholder planning capabilities, the ForestERA project prioritized the transfer of high-quality datasets and analytical tools within a framework of forest landscape ecology. To this end, we created a spatial decision-support system (Figure 4.2) and GIS tools for developing and comparing alternative forest restoration plans. It became apparent while assessing stakeholder needs that collaborative, participatory planning would be essential in formulating socially viable landscape-scale restoration strategies. To meet this need, ForestERA developed a set of collaborative planning processes that were used in conjunction with the spatial decision-support system to support stakeholders in clarifying their restoration priorities.

4.3.1. Educating Stakeholders on the Application and Utility of Analyses Based on Forest Landscape Ecology

A key component of our project has been to educate stakeholders about concepts such as the importance of planning at landscape scales. We have sought to develop a landscape perspective among the key stakeholders, allowing subsequent interactions to focus more meaningfully on big-picture issues related to the cumulative effects of multiple, independent management decisions and the spatial patterning of fire events, wildlife habitats, and human communities. Because we are developing and acquiring data and tools to support people outside our group in formulating management recommendations, it is important that they understand at least the basics of our models and assumptions. The educational component of our work has presented many forest landscape ecological concepts and methods to audiences of varying expertise. For example, we have described image-processing techniques, ecological modeling methods (such as those used to develop the wildlife habitat and fire behavior models), and spatial analyses (such as cumulative effects, noise filtering, and map projections).

4.3.2. The ForestERA Spatial Decision-Support System

The ForestERA spatial decision-support system allows users, both individually and collaboratively, to develop and compare alternative management action plans prioritized by location. Its design lets stakeholders define objectives for each scenario based on their preferences, apply criteria for designing and prioritizing management actions, and explore the trade-offs between alternative strategies. The first step (Figure 4.2) involves defining a decision problem pertinent to current forest conditions (e.g., inadequate protection of human communities and resources against high-intensity crown fires) and identifying goals for addressing the problem (e.g., reduce fire hazard in and around human communities and resources while minimizing negative impacts on wildlife). Part of this first step requires stakeholders to identify specific management objectives for meeting the broader goals of the scenario. In the second step, stakeholders define evaluation criteria for measuring the degree to

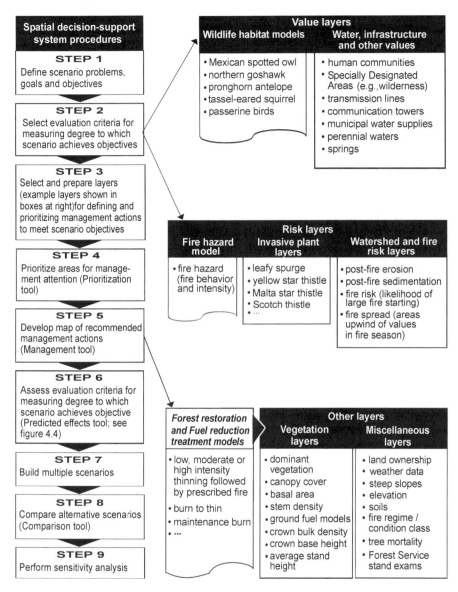

Figure 4.2. Schematic diagram of the ForestERA spatial decision-support system, showing the relationships between the system components and data layers. After identifying objectives and evaluation criteria, stakeholders select layers that represent the values and risks and use these layers to identify priority areas (step 2) and management actions (step 5). Forest restoration models are used to adjust the vegetation layers, which allow evaluation criteria such as the fire hazard and wildlife habitat characteristics (shown in bold) to be estimated within user-selected areas.

which a scenario achieves these objectives. For example, two objectives for the overall goal in this example could be to reduce the fire hazard to communities and the Mexican spotted owl (*Strix occidentalis lucida*), and the outcomes could be measured in terms of the reduction in potential heat output from fires a defined distance upwind of communities and Mexican spotted owl protected-activity centers. A protected-activity center is a regulated habitat category established around nesting areas of resident owls (as defined in USFWS 1995). Next, geographical data are chosen for use in defining and prioritizing management actions to meet management objectives. These data generally fit into the categories of *risks* such as the predicted intensity of a fire or the potential for soil erosion following stand-replacing wildfire, and *values* such as the community infrastructure and the characteristics of wildlife habitat. Users can select spatial data developed or provided by ForestERA, as well as their own layers. Criteria for unavailable geographic data are noted, and these data gaps are included in the presentation of the scenario results. During the ForestERA project, we developed data layers for fire risk (Dickson et al. 2006), watershed risks, vegetation composition and structure (Xu et al. 2006), characteristics of wildlife habitat (Prather et al. 2006), and fire behavior (developed using the FlamMap fire behavior program; Finney, in preparation). We provide each layer with an estimated uncertainty.

Following selection of the data layers that will be used to prioritize management actions, each layer is normalized by converting the values to a range between 0 and 1 so that the scales of each value are comparable before combining the layers in an overlay that permits simultaneous analysis of all layers. Stakeholders then assign weighting factors to the prioritization layers based on their preferences and available data. For example, a stakeholder might assign higher importance (greater weight) to areas upwind of developed areas than to the surrounding forest so as to address an objective based on protecting communities from fire. Stakeholders have the option of specifying jurisdictional, ecological, political, or other constraints for limiting the priority and location of modeled treatments. The weighted layers that represent values (map A in Figure 4.3) and risks (map B in Figure 4.3) are then combined (map C in Figure 4.3), thereby generating a layer that presents the priority of each area for management attention.

Stakeholders then use the ForestERA tools and spatial data to develop rules for locating management actions within the landscape based on the scenario's management objectives (for example, maps D to F in Figure 4.3). In this figure, maps A to C and management recommendations D to F were developed separately by different stakeholder groups (designated as "Green" and "Yellow") that participated in the 2004 Western Mogollon Plateau Adaptive Landscape Assessment process, and partial results are presented here to illustrate some key points in the process. Sisk et al. (2004; volume II) provide a complete list of the management recommendations that resulted from the May 2004 Assessment. Users can select from a suite of treatment definitions (e.g., low-intensity thinning followed by prescribed burning) or can define their own. Because the current debate on restoration and fuels management in southwestern forests centers on restoring ecosystem health and function,

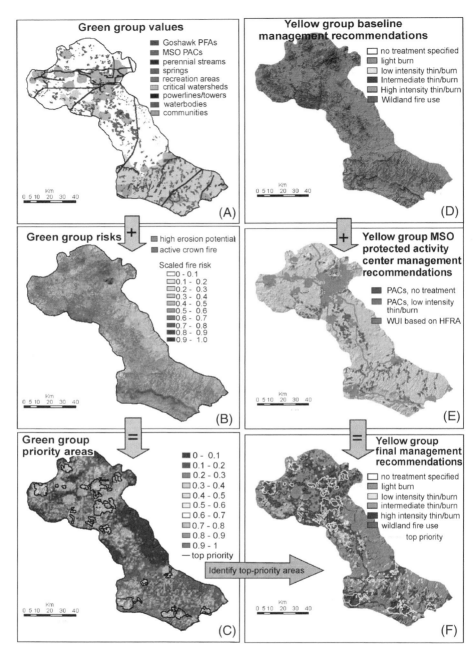

Figure 4.3. Two parallel processes for using (A) values and (B) risks to (C) prioritize areas and specify treatments (D and E) are combined to develop a sample management action scenario (F). The Green group was one of four breakout groups at the first Western Mogollon Plateau Adaptive Landscape Assessment workshop, held in February 2004, who developed prioritization maps. The Yellow group, made up of different stakeholders, was one of four breakout groups at the second workshop, held in May 2004, who developed management recommendations. HFRA, Healthy Forests Restoration Act of 2003; MSO, Mexican spotted owl; PACs, protected-activity centers; PFA, postfledging areas; WUI, wildland–urban interface (*See Colour Plates between pages 132–133.*).

stakeholders did not request that commercial logging be included among the management actions; retaining large trees and old-growth forest were more pressing objectives. For example, the Yellow team in one of our workshops chose to apply "baseline" treatments based on vegetation type and predicted fire behavior such that areas predicted to burn with high intensity were assigned more intensive treatments (map D in Figure 4.3). Users have the option of defining spatial constraints on treatments, consistent with those they applied when prioritizing areas, to provide details on application (e.g., a higher treatment cost on slopes between 40 and 70%) or to restrict management actions at certain locations (e.g., in map E of Figure 4.3, the Yellow team excluded protected-activity centers for the Mexican spotted owl outside of developed areas from being treated).

Once management action scenarios have been developed (e.g., map F in Figure 4.3), the ForestERA tools can be used to predict the effects (Hampton et al. 2003) of the treatment alternatives on selected evaluation criteria, such as the characteristics of wildlife habitat (Prather et al. 2005), fire hazard, and other parameters relevant to fire and forest ecology. For example, stakeholders may use the tools to predict changes in stem density and fire hazard surrounding human infrastructures, habitat suitability for pronghorn antelope (*Antilocapra americana*), and species richness of passerine birds following certain management actions within certain areas or the full study area. The model calculates changes in fire hazard and wildlife criteria by estimating the changes in forest structure (e.g., tree density) that would result from the management actions and using those changes as inputs for fire and wildlife models. Upon reviewing the results of this modeling in terms of the evaluation criteria (i.e., the predicted effects of each treatment scenario), stakeholders may develop new scenarios by adjusting the weighting factors, selecting additional criteria to include in the analysis, or changing the assumptions. Through this process of updating scenarios, stakeholders develop a suite of management recommendations and compare them in a number of ways, such as by graphing the predicted effects, compiling tables of areas affected by each type of action, creating summary maps that show the level of agreement between scenarios, or compiling a decision matrix (e.g., Table 4.1).

The decision matrix (Malczewski 1999) is a powerful multicriteria decision-support tool for exploring trade-offs between alternatives. It allows users to develop a single numerical score for each management scenario based on user-specified evaluation criteria. For example, the grayed values in Table 4.1 highlight the scenarios with the highest score for each set of weighting factors based on the specified preferences for objectives such as community protection. Specifying these preferences quantitatively has the added benefit of clarifying assumptions that may not otherwise be stated or obvious.

The final step is to perform a sensitivity analysis that tests the robustness of the ranking of alternatives. When relatively small changes in the data and in the stakeholder's preferences do not change the ranking of the results, a robust solution has been attained; otherwise, additional analyses such as boosting the precision of certain model components may be necessary. Using sensitivity analysis clarifies how the decision elements interact and provides stakeholders with a deeper understanding of the trade-offs involved in meeting their objectives.

Table 4.1. A sample decision matrix (adapted from Hampton et al. 2003) based on three hypothetical management plans (scenarios 1 to 3)

	Predicted effects across study area (%)			Weighting factors applied to predicted effects for each evaluation criterion	
				extent to which the scenario achieved the objectives	
Evaluation criteria	Scenario 1 Owls not considered in treatment placement	Scenario 2 Reduced treatment intensity in owl habitat	Scenario 3 Treatments excluded from owl habitat	Evaluation 1 Community fire protection and mitigating short-term impacts on wildlife	Evaluation 2 Community fire protection only
Reduction in fire hazard (kJ/m^2)					
Upwind of urban areas	9.3	8.3	9.8	50	100
Not upwind of urban areas	3.5	0.2	0.2	0	0
Habitat of the Mexican spotted owl (ha)	−12.3	−10.4	−0.1	25	0
Pronghorn habitat quality (dimensionless)	5.7	4.3	5.6	15	0
Squirrel density (squirrels/ha)	−2.4	−2.1	−2.4	5	0
Wood-pewee, *Contopus sordidulus* (probability of detection)	−4.0	−1.3	−8.0	5	0
Evaluation 1 score: Evaluates the scenario in terms of how well it protects communities against fire and protects or ameliorates wildlife habitat	2.1	2.0	5.2		
Evaluation 2 score: Evaluates the scenario only in terms of how well it protects communities against fire	9.3	8.3	9.8		

Grayed values represent the highest total scores for each restoration plan (scenario) based on the weighted predicted effects for each evaluation criterion. Note that Scenario 3, in which habitat of the Mexican spotted owl is not treated, is the preferred scenario (i.e., it has the highest score) when evaluated in terms of the extent to which the scenario either protects communities from fire alone (evaluation 2 score) or protects both communities and wildlife. In this study area, excluding owl habitat from treatment actually shifted treatments from areas with lower fire hazard that were not upwind of human communities to forests that were upwind, thereby better achieving both fire and wildlife objectives. This type of assessment can help stakeholders find common ground for reaching agreement on management plans or can aid in designing other potentially acceptable scenarios.

Supporting Public Participation

We have developed a suite of software tools and a user manual to assist stakeholders in carrying out the spatial analysis that we described in this section. Our prioritization tool guides stakeholders through the process of selecting and overlaying spatial data to generate a map of priority areas that can be used to rank treatment areas or select the highest-priority areas to be managed within a given time period (e.g., a 5-year plan). We have also provided a filtering tool that helps users to identify the highest-priority areas within the assessment area that are larger than a specified minimum size. We used the filtering tool to select the top-priority areas (outlined in black in map C and yellow in map F of Figure 4.3) based on an estimate of the total area that could be treated within 10 to 15 years (i.e., 180 000 acres = 73 000 ha) across the entire assessment area. The management action tool allows stakeholders to build alternative scenarios (maps of treatments and other management actions) based on specified rules for each spatial layer.

Another tool (Figure 4.4) predicts the effects of management actions on forest structure, the characteristics of wildlife habitat, and fire behavior. This information is intended as a guide for building additional management scenarios that better meet the user's objectives. For example, when the short-term negative impacts on habitat suitability for the Mexican spotted owl are higher than desired, treatments could be moved elsewhere or changed to mitigate these impacts. Changes in other evaluation criteria, such as fire hazard, can be assessed at the same time.

4.3.3. The Value and Function of Collaboration during Decisionmaking

In addition to technical datasets and tools, the ForestERA project developed collaborative, science-based planning approaches that combine the efforts of the planning team and other partners. That is, we have gone beyond the development of a spatial decision-support system to create a *collaborative process* that facilitates the use of this tool by diverse groups of stakeholders. Resolution of complex, interdisciplinary issues with ambiguous solutions that are subject to methodological criticism requires the inclusion of a range of nonscientist participants and perspectives (Funtowicz and Ravetz 1995). These participants may serve several purposes, both as experts and as members of society. They may represent a "community of inquirers" with a deeper understanding of the problems and the available solutions (Norton 1998). This community may be a source of local knowledge, and may help to define the most relevant policy problems. It may also serve in a review capacity, providing a critique of the quality of the data and the assumptions used in the analyses and participating in more traditional research that focuses on individual disciplines (Cortner and Moote 1999). Finally, it offers an alternative to technocratic decisionmaking and may thus prevent purposeful or inadvertent omission of relevant information from nontraditional sources and increase the probability of obtaining widespread support for the outcome of the problem-solving process (Glasser 1995).

We have designed our spatial decision-support system to assist and promote a civic approach to science (Cortner 2003) in which citizens participate directly in a process designed to promote collaborative learning and democratic deliberation.

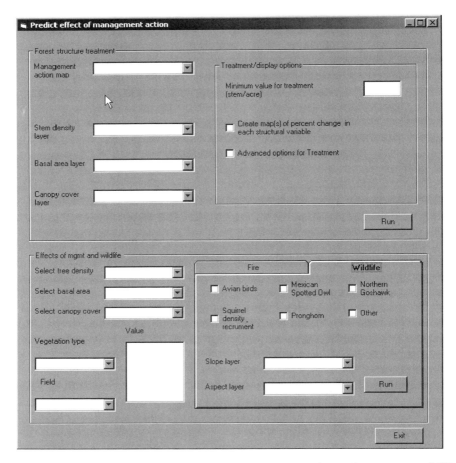

Figure 4.4. The ForestERA tool for predicting effects of management actions on forest structure, wildlife habitat characteristics, and fire behavior. This tool has been developed as an extension of the ArcGIS software (ESRI, Redlands, CA).

Harris (1995) describes the need for scientists, resource managers, and the public to learn from each other so as to effectively bring science and policy together. Typically, management of southwestern forests has been planned by experts on a project-by-project basis for areas ranging from 5000 to 50 000 acres (2000 to 20 000 ha), with only minor up-front input from other interested stakeholders who focus on the cumulative effects of these projects. We are seeking to strengthen the interest and ability of diverse stakeholders to engage in and contribute to forest restoration planning, while at the same time providing tools that will allow them to develop and prioritize alternative management plans at broader spatial scales as large as millions of acres (Sisk et al. 2006).

The management of forests and other public resources has long been guided by attention to the public good and the principle of sustainable management. In 1905, Gifford Pinchot, who would become the first chief of the U.S. Forest Service, defined the objective of forestry as the effort "to make the forest render its best service to man in such a way as to increase rather than to diminish its usefulness in the future" (Pinchot 1905). Subsequent development of the federal bureaucracy enshrined this principle in a management model that privileged professionalism and science. The National Forest Management Act of 1976 formalized this approach by calling for periodic assessments of forest lands, the implementation of management programs based on multiple-use and sustained-yield principles, and the development of resource management plans for all public forests. Coupled with the National Environmental Policy Act of 1969, this Act defined a role for public participation in policy development, and prescribed a process of public review and comment that should be implemented "early and often throughout the development of plans."

Both Acts have stimulated unprecedented public involvement in federal decisions that affect public forest lands. In some cases, litigation resulting from this involvement has stopped planned activities and reversed management decisions. In others, it has slowed forest planning and management activities to a crawl, though the frequency of this litigious "gridlock" is hotly debated. Whatever the cause, forest management in the United States has become a highly polarized issue, with heated public exchanges and prolonged policy battles framing increasingly difficult issues, many of them revolving around cutting of trees, fire management, and the conservation of habitat for sensitive species. In this politicized context, rational scientific approaches often fail to sway the policy debate (Sarewitz 1996). Scientific uncertainty, specialized terminology, and structural impediments in the policy process often minimize the influence of scientific input in policy debates (Lindblom and Woodhouse 1992).

In these situations, science and technology can be marginalized to the point that representatives from different sides of the debate (for example, those who favor aggressive thinning of overstocked forests and those who oppose entry of machines into roadless areas) may selectively screen the available scientific information to bolster preexisting policy positions rather than drawing on the full body of scientific knowledge to develop the most effective policy (Porritt 1994). In such a polarized environment, science may have little to offer. In fact, it may make matters worse by focusing attention on issues characterized by high uncertainty and, potentially, confusing the debate and deepening the disputes (Sarewitz 2004). In such situations, knowledge transfer becomes a broader task that not only encompasses the technical and scientific elements that inform resource management, but also delivers robust and flexible tools that educate the stakeholders and that help these stakeholders include all the available information in making their decisions. In the ForestERA project, we set out to integrate the concepts of encouraging public participation in science ("participatory science"; Fischer 2000) and public input into the debate that characterizes the democratic process ("discursive democracy"; Dryzak 1990) so that the public, the experts, and the decisionmakers could all draw on a common set of reliable data and use shared analytical tools to explore possible management scenarios and, in so doing, rebuild and strengthen the public discourse that empowers democratic approaches to the management of public lands.

We have supported the use of our forest landscape ecological tools in several collaborative planning venues, two of which we describe in the next section. By providing easy-to-use tools and by developing workshop-based processes that allow stakeholders to explore the predicted trade-offs in alternative plans, we are facilitating the adoption of landscape-scale analysis in forest planning. Our success in developing useful tools has been tied to the fact that before and during the development of our data and tools, we have dedicated significant time and resources to outreach activities designed to educate and encourage the involvement of stakeholders—an essential component in transferring forest landscape ecological knowledge and technology.

4.4. KNOWLEDGE TRANSFER MECHANISMS

4.4.1. Engaging Stakeholders in Project Development

Given the ForestERA project's prioritization of participatory and accessible science, outreach and collaboration with a diverse array of stakeholders during project development has been critically important. We initially identified 21 stakeholders, ranging from federal and state agencies to local groups and litigation-oriented nongovernmental organizations. Stakeholders were chosen according to their historical participation in regional projects related to forest restoration so as to adequately represent a diversity of constituencies and political positions in the ongoing debate.

During the initial project-planning phase, we assessed the needs of these groups through personal interviews, surveys, and a workshop. We then summarized our findings for all participants. Identification of stakeholder needs helped to guide project planning in a transparent manner, prioritize goals and deliverables, and increase the likelihood that the deliverables would be relevant and valuable to a wide array of stakeholders and easily integrated into the ongoing debate in the Southwest. In a broader sense, initial outreach began as a discussion with some stakeholders about their values and priorities related to landscape-scale restoration planning. Stakeholders helped to design the common source of information, and were given a voice in determining the means by which science could be used to inform decision-making and policy development.

After our initial communication with the stakeholders, open invitations for collaborative research were offered to all stakeholders. By inviting active collaboration, ForestERA staff were able to build on the decades of research that has led to our current understanding of the condition of the forest ecosystem and its restoration needs in the Southwest. This has also allowed ForestERA to play a valuable role in integrating research results from different disciplines, organizations, and individuals into a common decision-support framework, thereby encouraging the use of science in ongoing policymaking processes.

Communication with stakeholders has occurred throughout the project and has been critical in fostering discourse about the planning of landscape-scale restoration. Periodic project updates and reports have informed stakeholders about progress, challenges, and new opportunities related to the development of datasets and tools. E-mail updates and Web site development have enabled stakeholders to follow the project's progress. Several presentations to stakeholders have served similar purposes and, more importantly, have facilitated two-way communication between stakeholders and the ForestERA team. After a year of data and tools development, we surveyed stakeholders to obtain feedback on our progress, and refined future project goals by reassessing stakeholder priorities and their critical information needs. Finally, in order to maintain connections within the academic community, ForestERA staff have presented progress reports and preliminary results at a number of conferences (e.g., Sisk et al. 2005; Xu et al. 2005), and have submitted manuscripts to a variety of management-oriented and academic journals.

4.4.2. Delivering Data and Tools Designed to Meet Stakeholder Needs

As one of the primary goals of the ForestERA project has been to build stakeholder capacity to identify and assess landscape-scale restoration opportunities, the distribution of datasets and tools to a wide variety of stakeholders has been of paramount importance throughout the project. We used several stakeholder surveys to prioritize the development of data and tools such as spatial layers for vegetation structure and fire hazard that span multiple jurisdictions. Most stakeholder needs could be met reasonably well, but some (such as the provision of Web-based tools for exploring and analyzing datasets) could not be met given time and other constraints. Based on the survey results and other information, we developed or acquired datasets that we delivered at no charge via the ForestERA Web site (http://www.forestera.nau.edu/) and on compact disk, complete with detailed descriptions, rigorous accuracy assessments, and other information about the data ("metadata"). As many stakeholders indicated a preference for tools compatible with their existing GIS software, and specifically with ArcGIS (ESRI, Redlands, CA), we developed an ArcGIS extension that steps users through our spatial decision-support system and an accompanying user manual. We have completed one training session for users of the tools and are planning additional sessions for stakeholders throughout our study area.

4.4.3. Letting the Stakeholders Drive: Testing Forest Landscape Ecological Tools during Their Development

Assessing stakeholder needs during the preliminary phases of project development and communicating and collaborating with stakeholders throughout the development process were essential in building support for the approach and the technology during the project's latter stages.

However, these processes were not sufficient to design our tools or prepare stakeholders to use new datasets and analytical approaches in collaborative planning.

To engage stakeholders in using the ForestERA tools collaboratively, and to strengthen the link between the science developed within the ForestERA project and critical social and management concerns, we invited six key stakeholders to participate in a pilot process (Hampton et al. 2005) aimed at applying and testing ForestERA's capabilities in real-world planning. Although we were concerned that our tools were not ready for rigorous use, the experience and information we gained from helping stakeholders to develop planning scenarios that reflected *their own* preference*s* proved invaluable.

The six participants in the pilot process came from diverse constituencies: USDA Forest Service management staff from the Coconino National Forest and research staff from the Rocky Mountain Research Station, three nongovernmental organizations (the Grand Canyon Trust, the Center for Biological Diversity, and the Greater Flagstaff Forests Partnership), and academic researchers from the Ecological Restoration Institute. During facilitated focus-group discussions, the entire group became familiar with the opportunities and limitations related to landscape-scale planning. Many uncertainties inherent to the development of datasets and the model were discussed, and the need for sustained discussion to address these uncertainties became apparent. Overall, the focus-group testing built a group of participants capable of and willing to continue working in a collaborative environment to address landscape-scale forest restoration challenges, and provided concrete and practical feedback for refining our tools.

4.4.4. Transfer of Collaborative Planning Processes

By supporting collaborative planning with our data and tools, we have engaged a wide variety of stakeholders in discussions surrounding landscape-scale restoration, and have continued to build stakeholder ownership in the ForestERA project. We will discuss two ForestERA-supported collaborative planning processes in this section: the Western Mogollon Plateau Adaptive Landscape Assessment process and the Community Wildfire Protection Planning process.

In August 2003, the Ecological Restoration Institute received funding from the USDA Forest Service to assist the Coconino National Forest in expanding forest health restoration and fuel break treatments. The Institute was funded to develop a collaborative, adaptive approach to assessment that would facilitate landscape-scale treatment and planning; in this context, *adaptive* refers to the option of refining planning strategies based on data collected using monitoring protocols (ERI 2004). The Institute proposed a facilitated workshop, using the ForestERA system for scenario analysis to assess restoration priorities. The ForestERA project began working with the Institute in September 2003 to design a workshop-based planning process in which policymakers, technical support groups, and other stakeholders could use the project's datasets and tools, in addition to other existing datasets and tools, to develop scientifically rigorous, socially viable, restoration strategies. This process became known as the Western Mogollon Plateau Adaptive Landscape Assessment process, and included two workshops and two Web-based workshops. For more details, see ERI (2004) and Sisk et al. (2004).

The ForestERA project has also used its datasets, tools, and planning approaches to support the development of Community Wildfire Protection Plans, as mandated by the federal Healthy Forests Restoration Act of 2003. During the summer and fall of 2004, ForestERA staff supported the development of the Greater Flagstaff Area Community Wildfire Protection Plan (GFFP and PFAC 2004) and the Rim Country Community Wildfire Protection Plan (GCA 2004), both in northern Arizona. Within both processes, planners used ForestERA data and planning approaches to identify high-priority areas for protecting communities and appropriate management activities within those areas.

Fostering stakeholder ownership of workshop outcomes was a high priority during the development of the Western Mogollon Plateau Adaptive Landscape Assessment and two Community Wildfire Protection Plans. Several strategies helped us to provide this ownership. Prior to the first Assessment workshop, in February 2003, key stakeholders were chosen to serve on a planning team that would define the goals, objectives, and strategies of the Assessment for engaging stakeholders in the planning process. Engaging this team, including four of the six participants in the pilot process, increased initial participation in the process and helped to recruit ambassadors who could assist in interpreting the datasets and planning approaches for diverse constituencies. Within the Community Wildfire Protection Plan planning processes, similarly diverse planning teams convened to discuss appropriate protection strategies.

We surveyed participants prior to the workshops to identify their greatest concerns and any additional data needs so we could design a highly relevant and valuable process. During development of the Western Mogollon Plateau Adaptive Landscape Assessment and the Community Wildfire Protection Plans, participants were encouraged to think beyond the ForestERA datasets and tools. They were encouraged to use ForestERA datasets and tools in combination with other sources of information to develop appropriate recommendations. Participants were also encouraged to annotate any datasets that they used. These comments were recorded and incorporated, when appropriate, in the final plans and workshop reports. In the end, modified versions of the ForestERA datasets were accepted and used by most participants, even though participants were not forced to use these tools. In developing and using new data, analyses, and planning approaches in an environment characterized by proprietary attitudes and "turf" issues, this kind of "soft touch" has been extremely useful in minimizing resistance to the new information and approaches.

Because the ForestERA approach cannot dictate the modes of discourse that will surround the planning process, we focused on the ability to integrate science into decisionmaking in a variety of political environments beyond the collaborative processes summarized above. Although transparency, pluralism, and discussion are vital in landscape-scale planning, there is no guarantee that planning and decisionmaking will embrace these characteristics. As such, the ForestERA project is strengthening our relationships with stakeholders likely to participate in collaborative processes, and is designed to support such discourse rather than developing tools that are only useful when such discourse occurs. The ForestERA spatial decision-support system was designed for and is being distributed to individual stakeholders and stakeholder groups to help them integrate rigorous science in their decisionmaking, planning processes, and stakeholder review of land management plans.

4.4.5. Supporting the Learning Process Using ForestERA Products

As explained in Section 3, our work requires us to educate our stakeholders in the concepts and methods of forest landscape ecology, such as modeling wildlife habitat characteristics, and on the appropriate use of these tools, for example by validating models and metadata. Our strategy has been to present the material as simply and clearly as possible to nonexperts, while making the technical details available to those who can use them. In addition, we have designed our collaborative processes to encourage the sharing of knowledge among experts in fire and watershed management, public policy, and other areas. Besides facilitating these discussions, our educational support takes the form of recording the recommendations of these experts for distribution by means of reports and the Internet.

Identifying and assessing landscape-scale restoration opportunities using spatial modeling tools is a new process for most stakeholders in northern Arizona. Uncertainty about the highest priorities and the potential outcomes characterizes the ongoing discussions and planning processes. No restoration strategy will satisfy all stakeholders, nor can any strategy achieve all restoration goals. For these reasons, the ForestERA project has attempted to offer stakeholders opportunities for dynamic, adaptive learning. This learning process will lead to more informed discourse and the identification of scientifically rigorous, ecologically appropriate, and broadly acceptable strategies.

To integrate interdisciplinary perspectives into environmental decisionmaking and policymaking, we have encouraged participation by a diverse set of stakeholders. Because the interplay among these perspectives is complex and the means for formulating scientifically rigorous, socially acceptable strategies may not be obvious, we have adopted a descriptive, participatory analytical framework to support a larger goal of learning (Glasser 1995; Holling 1978; Lee 1993; Walters 1986). We have developed scientifically rigorous information and tools to support planning in the Southwest that remain understandable to a broad cross section of participants. Participants are supported in clearly stating their assumptions, asking "what if?" questions, reformulating their questions, and generating site-specific information that illuminates viable strategies.

To support learning during the planning process, ForestERA staff developed datasets that are accessible via slideshow presentations, the ForestERA Web site, and printed data atlases that are distributed prior to and during planning sessions. During these sessions, we support participants in developing unique restoration scenarios in breakout sessions, then compare the scenario assumptions and results during plenary sessions. By comparing these scenarios, participants can use creative problem-solving approaches and learn from others, while building ownership in the workshop outcomes. Following the planning sessions, scenarios were analyzed, refined, and presented for review by workshop participants. Within the Western Mogollon Plateau Adaptive Landscape Assessment process, Web-based meetings after the planning workshops allowed participants to further explore the scenarios they had developed. By facilitating this online discussion, we encouraged workshop participants to continue learning as one means of building ownership in the workshop outcomes and encouraged more stakeholders to join the discussion. Participation in these workshops was moderate, perhaps due to "technology fatigue" or competing demands on their time once back at their usual working environments. Nonetheless, archiving the workshop results online contributes to knowledge transfer efforts.

Through collaborative development, pilot testing of datasets and tools with real-world issues and stakeholders, and workshops aimed at building ownership in and understanding of the ForestERA project, we are increasing the ability of stakeholders to perform landscape-level analysis. Demand for planning based on this analysis is increasing in our study region. Requests for ForestERA datasets and tools by federal agencies, nongovernmental organizations, and litigious environmental groups alike are increasing steadily. We hope that sustained efforts to transfer the ForestERA approach will shape the tone, tenor, and scope of future planning efforts, and most importantly, will promote the restoration of forest ecosystems in the Southwest.

4.5. SUCCESSES AND CHALLENGES IN KNOWLEDGE TRANSFER

Most natural resource issues involve public lands and, thus, public input, and scientists often comment on the policy that results from such processes. Yet science-driven approaches to planning are often inaccessible to nonscientists, and many are never implemented because they fail to account for how policy is developed. This disconnect has undermined many attempts to apply knowledge of forest landscape ecology, and modern science in general, in environmental management. There is an obvious need for transparent, accessible, and rigorous scientific approaches that are also accessible to nonscientists so these citizens can participate more fully in the planning process.

4.5.1. Successful Use of the ForestERA Data and Tools

The ForestERA data and tools were used successfully to produce collaboratively developed recommendations and assessments for forest planning in the Western Mogollon Plateau Adaptive Landscape Assessment and two Community Wildfire Protection Plans. Participants in the Flagstaff plan took the results one step further by using maps of the management recommendations to estimate treatment costs for both high-priority areas and the entire landscape. In addition, representatives from the Apache–Sitgreaves, Coconino, and Kaibab National Forests, and from the USDA Forest Service Southwest Regional Office, have identified several potential uses of ForestERA tools: (1) to help plan and prioritize future management, (2) to evaluate and validate assessments of hazard and risk, (3) to inform analyses under the National Environmental Policy Act in future projects, (4) to help the National Forests compete for funding, and (5) to inform revisions to Fire Management Plans and Forest Plans. There has also been considerable interest in using our spatial datasets in projects not directly related to our work.

In addition to supporting specific planning forums, a host of less-tangible benefits have arisen from the ForestERA tools and science-based collaborative process. These include empowering stakeholders, strengthening relationships, and raising awareness of consensus values. The sharing of ideas among workshop participants is an important benefit that improves their understanding of different points of view. In our workshops, participants moved beyond discussing values and risks conceptually to reviewing actual maps of these factors. Participants could then discuss the actual locations at highest risk from wildfire or other threats. Providing concrete examples of how multiple values interact allowed stakeholders to deepen their understanding of the area of interest and thereby make more informed recommendations. Our work can thus serve as a model for other planning areas, and we invite readers to contact us to learn more.

4.5.2. Overcoming Obstacles

Based on our experience, the obstacles to transferring knowledge of forest landscape ecology fall into four general categories:

- technical challenges related to data quality and abundance, modeling uncertainty, and changes in the capability and use of spatial software over time
- challenges in working with stakeholders, including scientists
- empowering the public without oversimplifying the situation
- organizational challenges (e.g., statutory constraints, a lack of time and other resources)

Technical challenges commonly arise related to data quality and abundance, modeling uncertainty, and changes in the capability and use of spatial software over time. To address the latter issue, we surveyed stakeholders to determine their depth of training and commitment to various GIS applications. After considering the possibility of developing stand-alone software, we decided to develop an extension to a commercial GIS application (ArcGIS, ESRI, Redlands, CA) that was already being used by most stakeholders. Changes in this software could render our tools less useful in the future, but we have found that it is more time-consuming and challenging to develop the models than to encode them and that relying on familiar software facilitates knowledge transfer.

Forest landscape ecological projects often lack geographical data at the multiple scales across which the system's dominant processes occur, particularly for important processes that are difficult to model with certainty, such as vegetation

Supporting Public Participation

recruitment. Stakeholders were aware of and frustrated by these issues, and directed us early on to build new and more accurate spatial layers for basic data, especially for current forest composition and structure. Although this effort required significant resources, we provided and continue to supply our stakeholders with new vegetation, fire, wildlife, and watershed data via our Web site. We cannot emphasize enough the appreciation and interest gained by offering these key products freely and with a guarantee of accuracy.

Providing high-demand data layers fosters appreciation of and facilitates the application of forest landscape ecological knowledge, but some desired data or models (e.g., population viability analysis) are unavailable. In addition, the demand for less-important layers increased after we delivered the highest-priority layers. We found that encouraging stakeholders to develop management plans with existing tools despite any data gaps prevented stakeholders from the "paralysis" that results from waiting for perfect data and analyses. Because science always generates more questions as new knowledge is added, it is imperative to focus on providing the good science that is available and avoid being distracted by unresolved questions. There is general agreement that adaptive management, in which assumptions and decisions are updated based on the results of past management, is a reasonable and effective way to move forward. After all, accepting the inaction that results from gridlock in planning is also a decision that has associated impacts.

Working with stakeholders, including scientists, represents the second category of challenges. During workshops, it is difficult to find a balance between *directing* collaborative activities and allowing participants to choose their own course, thereby fostering creativity and ownership. With too little leeway, participants feel that their creativity is ignored; with too much, groups can flounder around and fail to develop a framework for action. Our spatial decision-support system provides a sufficiently broad framework for group progress, and examples of decision problems and objectives can facilitate brainstorming without imposing our worldview on participants. It is also imperative that group discussion not be driven solely by hard data. Allowing stakeholders to talk conceptually about priorities and criteria for selecting management actions increases participation in the process.

Another challenge relates to empowering the public without oversimplifying the situation. We have addressed this issue by supplying information in a range of formats and levels of detail. In the data atlas that we distribute at workshops, we include descriptions of the data and their accuracy in simplified terms, but also provide more detailed metadata and use more technical language in manuscripts for more technically capable participants. Because it is difficult to ensure that the spatial data we distribute are used and interpreted appropriately, it is important to provide recommendations for proper use of the data.

In developing the ForestERA tools, conducting collaborative scientific analyses took additional time; however, the benefits (improved quality and avoidance of turf battles) made the extra effort worthwhile. A related issue involves avoiding confusion over work by other groups. To the extent possible, we have sought collaborations with these groups or have at least assessed the extent to which our work overlaps in time, space, and support. However, the current status of other projects and the inherent uncertainty in future funding limit the amount of joint planning that can be done.

Another challenge in using knowledge of forest landscape ecology in real-world forest planning arises primarily from the fact that these concepts are only just beginning to take root in many management agencies. Formidable barriers to use of this knowledge and related technologies include the high start-up costs of acquiring high-quality data and allocating employee time to spatial analysis training and work. Although most stakeholders want to shift from the current focus on project-level planning to include broader considerations, existing approaches to analysis can hinder progress. For example, government agencies must follow certain statutory guidelines for their management actions that require them to classify the landscape into specific categories that may be irrelevant for more general use or that may not allow the estimation of cumulative effects across larger landscapes. When we could not reach general agreement on the use of new techniques, we derived specific data layers that would be useful to specific stakeholders, such as a layer for the Mexican spotted owl that was tailored to the Recovery Plan used by the USDA Forest Service. This brings up the related point that in multijurisdictional projects, it is difficult to learn of all stakeholder planning efforts and understand how our tools can best be used in those efforts. We see this as a two-way process in which we tailor our projects to stakeholder needs, present our preliminary results, and rely on engaged stakeholders to provide the necessary feedback. However, it is difficult to find planners who have enough time and interest to adapt our tools to overcome institutional barriers such as their own workload and statutory constraints. The support of upper management can significantly reduce these barriers.

We have also been challenged to redefine the role of our project as we move from development and design of tools to their application. We see the role of our research group as piloting new forest landscape ecological methods and processes to encourage public participation in forest planning. And although we are providing scientific, technical, and public support to facilitate the use of our tools in planning, we hope to eventually transfer this role to the appropriate management agencies and cooperators. A related issue is that some stakeholders are reticent to adopt these methods when they are unsure who will maintain, update, and support ForestERA's tools in the future. Finally, there is the issue of determining an appropriate role for science and new information. Early on, we decided to provide tools that would support decisionmakers rather than making specific recommendations. However, this distinction is not always clear. For example, we have generated maps that show recommended levels of forest treatment to reduce fire hazard based on differences in forest structural attributes (e.g., crown bulk density). These could be misconstrued as policy recommendations, even though they were developed solely to support decisions.

4.5.3. Changes for Future Efforts

We are continuing to refine the collaborative processes we have developed as we embark on two additional projects in the Southwest, which are funded through 2006.

In each project, we are transferring knowledge of forest landscape ecology horizontally (into new regions) and vertically (by developing new data layers and models at varying scales). When we work in new regions, we must adapt our approach to meet different stakeholder needs. However, some aspects of our role remain constant, and we are beginning to plan how to transfer these components. As an academic research group, our strengths and interests focus on tool development and knowledge transfer rather than on supporting ongoing planning operations.

As the demand for ForestERA data and tools has increased, our transfer efforts have continued. For example, after the Western Mogollon Plateau Adaptive Landscape Assessment process, the Coconino National Forest asked us to hold a 2-day collaborative forum focusing on one Forest Service district within the larger area (Forest ERA and CNF 2005). Following this process, a district ranger asked us to train his staff to use our tools so they could develop alternative management scenarios for their 5-year fuels plan. The Ecological Restoration Institute conducted 10 interviews with participants in the Assessment to determine the effectiveness of the process (Abrams 2005a), which we found useful for project planning. To assess future projects, we are working with social scientists to conduct pre- and posttransfer assessments to evaluate how the tools met stakeholder needs and to better understand changes in participants' preferences following use of our spatial data and involvement in our collaborative processes.

We are continuously improving the technical aspects of our work, including methods to convey uncertainty and work at multiple scales; for example, we are focusing more on the lands surrounding our bioregion of interest. We are also improving the design of our collaborative assessments and are organizing training workshops. Simultaneously, we will continue to distribute data and transfer ForestERA tools and collaborative processes to interested stakeholders, including agencies at city, county, state, and tribal levels, conservation organizations, engaged citizens, and others. By sharing high-quality data and providing user-friendly and scientifically sound tools for analysis and display of data and scenarios, we will continue to support landscape analysis in the service of ecosystem restoration.

4.5.4. What Others Can Learn from Our Experience

Our most important recommendation for facilitating the transfer of forest landscape ecological knowledge and technology is to target the transfer activities at key points in the project while supporting outreach activities throughout the project. These key points include the start of the project (communicating project goals and assessing stakeholder needs), when preliminary products become available (to obtain additional feedback), during and directly after collaborative workshops, once workshop reports are available, and when final data layers and tools are available. This will also include hiring team members with good communication and outreach skills. More specifically, forest landscape ecological products should be designed based on a careful assessment of stakeholder needs, with intermediate products presented to obtain feedback and allow ongoing improvement of the products, while providing training and consulting to encourage adoption of the final products. In short, success requires focused preliminary planning and targeted effort by personnel who can communicate effectively with stakeholders. Although the cost of this strong focus on transfer activities has been significant, the benefits include improved product design, the development of long-term relationships with stakeholders, enhanced stakeholder ownership of the work, and increased job satisfaction within our own team because our services are valued.

Involving stakeholders in the design of forest landscape ecological data and tools early and often ensures that the resulting products effectively meet the highest-priority needs. Stakeholder-needs assessments that summarize the methods and results must also be distributed. These are invaluable references throughout the project because they directly link stakeholder needs with planning decisions, such as the geographical data and analyses that were chosen. Because conditions and user preferences shift over the course of multiyear projects, project staff must remain flexible so they can mold products in response to changing needs without losing focus on the key deliverables.

Providing stakeholders with full access to our methods, data, and tools encourages a sense of ownership that engages them more fully in our work. Offering a broad range of distribution levels, from raw GIS data layers and complete metadata to an atlas with data described in plain language, facilitates understanding among stakeholders who range from expert spatial analysts to concerned citizens. It is also critical to warn stakeholders of unavailable or inaccurate data layers so plans can be made for future development of these layers, if warranted.

Demonstrating how stakeholders can apply our tools in their particular organizations can engage stakeholders who might otherwise question the need to learn new techniques. This can be especially tricky when there is little institutional emphasis on landscape-scale analysis, but presenting examples of new capabilities (e.g., predicting the cumulative effects of forest management) along with all the requisite data for the lands being managed often uncovers possibilities that break the "business as usual" mind-set. At the other extreme, as more organizations use forest landscape ecological methods and products, they must understand what sets our work apart from similar projects. Finding opportunities for new, practical application of this approach is often an iterative process, and stakeholders must be sufficiently motivated to uncover and explore promising possibilities. To succeed, it is necessary to foster relationships with stakeholders who are enthusiastic about learning and applying forest landscape ecological methods, while staying alert to opportunities for creating new relationships.

To assist in knowledge transfer, we recommend hiring a communicator (a stakeholder liaison) with a solid foundation in the social, economic, and scientific issues relevant to the project. Their skills must include the ability to communicate technical concepts to a wide range of people, to be perceived as an objective source of information rather than someone promoting an agenda, and to be conscious of, but not trapped by, stakeholder histories and relationships. Given that forest landscape ecological principles offer a new way to plan, some stakeholders may be interested

Supporting Public Participation

in landscape analysis but feel unable to join in the process because they have no background and little exposure to this approach. An effective stakeholder liaison can bridge the knowledge gap by finding ways to connect with these people. This skill set does not necessarily overlap with that of a competent spatial analyst.

Because much of the controversy among stakeholders involves a lack of scientific information that can be used to compare alternative plans, including the cumulative effects of multiple projects, or as a result of distrust of this information, we have focused on developing data and tools to fill this void. To reap the benefits of regional planning, the efforts of the various land management agencies and other stakeholders must be coordinated to some extent. Relatively small environmental, academic, tribal, or other stakeholders often lack the resources to dedicate to this type of planning, but nonetheless have valuable local knowledge and expertise they can bring to the planning process. We have learned that to effectively transfer our technology, we must take a leading role in designing and supporting collaborative processes in which we guide stakeholders to work together, supported by our tools, to develop restoration plans. Our stakeholders have encouraged us to take on this role, as they view us as an objective participant and essential to the success of the collaboration. This result underscored the importance of our decision to *support* decisionmakers rather than making specific recommendations ourselves, and to avoid any impression of favoritism in working with stakeholders.

The development of treatment scenarios by multiple stakeholders in workshop settings rather than by individual stakeholders or organizations working in isolation has many benefits: (1) stronger relationships and trust between stakeholders, (2) improved mutual understanding of divergent points of view, (3) clear understanding of agreements and disagreements, (4) encouragement of creativity, (5) faster development of solutions, and (6) isolation of participants from other demands (e.g., their daily workload) so they can focus on forest landscape ecological issues. We believe that the long-term potential for encouraging landscape-level planning and reducing litigation far outweighs the up-front expense and time of workshop-based processes. However, a broad spectrum of stakeholders must participate to ensure that the results of the process are not seen as biased toward certain interests. We recommend sending out invitations widely and early to promote strong participation.

Despite the multitude of benefits, fast-paced collaborative environments do not supplant the need for focused training in forest landscape ecology if long-term change is to be achieved. It takes time for stakeholders to fully digest and become comfortable with new information represented in multiple spatial layers and to understand the inner workings of various spatial tools. Multiple exposure to forest landscape ecological concepts and hands-on training increase their comfort level in using new data and decision-support tools.

Some tasks, such as assessing the feasibility of thinning forests on steep slopes, are more technical and specialized than others and can be greatly aided by expert guidance. However, the role of experts and of our project should be to *inform* stakeholders rather than making decisions for them. Although we have designed our forest landscape ecological products for use by both experts and nonexperts, working collaboratively, the benefits of greater public involvement are numerous, and are likely to result in longer-lasting decisions grounded in the best-available science.

In the Western Mogollon Plateau Adaptive Landscape Assessment workshops, our facilitator set a tone conducive to collaboration by defining ground rules that encouraged courteous and effective group dialog. We found that in this environment, participants were less afraid to tackle contentious issues than we had originally predicted. Minimizing the emphasis on organizational affiliations encouraged some stakeholders to contribute more creatively to group exercises because they were no longer intimidated by the feeling that they were representing their entire organization. Instead, they felt more comfortable acting as individual contributors to a team working toward shared objectives. Engaging participants by asking them to volunteer to perform services such as note-taking and reporting the results of breakout group discussions to the larger group encourages involvement in the process and ownership of the results, as well as reducing the workload on the workshop organizers.

In designing the collaborative workshops, consider partnering with successfully operating collaborative groups that are already involved in similar issues. When none are available, form a diverse planning team composed mainly of local stakeholders and clearly state what decisions you want them to make. For example, if funding for the workshop dictates that certain objectives must be met, then clearly identify these objectives at the outset to avoid misunderstandings or perceptions that you are manipulating the workshop process after it begins.

Establish methods for dealing with contentious issues such as missing data or developing a mutually acceptable definition of fuzzy concepts such as the wildland–urban interface. Try to emphasize the concept of adaptive management, as this is an effective way to move forward despite gaps in the data and incomplete knowledge of the consequences of certain actions. When participants question the quality of newly developed spatial data, explain why the data were developed, describe the accuracy assessments you performed, and provide any older or less-comprehensive data requested by stakeholders. Outdated, inaccurate, spatially incomplete, or otherwise undesirable data are still useful if stakeholders are accustomed to using them or know that they are available. Once the inadequacies of these data become clear, interest in them usually decreases and planning can move forward.

4.6. CONCLUDING REMARKS

ForestERA represents an appropriate model for transferring scientific understanding in a manner that informs public policy debates and provides practical tools for assessment and planning. Our primary goal has been to make forest landscape ecological principles and tools maximally accessible to a wide cross section of stakeholders in forest restoration. Our spatial data layers and models have been designed with input from these stakeholders, then subjected to accuracy assessment and independent evaluation. Real-world applications of our modeling tools have fostered informed discourse and broad participation in identifying management priorities. Alternative restoration scenarios have been developed based on explicitly stated objectives, and their effects on fire behavior and the characteristics of wildlife habitat have been evaluated and compared. ForestERA tools for iteratively developing forest management scenarios have deepened stakeholder understanding of the relationships among values, risks, and other decision criteria. The results achieved thus far reveal better-informed planning and closer collaboration among stakeholders at all levels.

When this chapter was written, we were continuing to support ongoing Community Wildfire Protection Plans and other planning efforts in the Western Mogollon Plateau, while developing and acquiring spatial data for similar projects in eastern Arizona and north-central New Mexico for which our community liaisons have completed needs assessments (Abrams 2005b, Schumann 2005). These and other efforts will explore in more detail how the ForestERA approach can support integrated planning for and management of lands managed by various entities, including tribes, state and federal agencies, and private landowners.

Acknowledgments

We are grateful to the many stakeholders who attended our open houses, filled out questionnaires, participated in workshops, and otherwise shaped this project with their diverse perspectives and experience. Special thanks go to Mark Finney and Chuck McHugh of the USDA Forest Service, Pete Fulé and Doc Smith at the Ecological Restoration Institute, John Bailey of Northern Arizona University, and Bill Romme of Colorado State University for their aid in modeling fires and forest treatments. We greatly appreciated the assistance of Brad Piehl of JW Associates, Inc. in watershed modeling. For wildlife modeling, we thank Norris Dodd, Mike Ingraldi, Steve Rosenstock, and Brian Wakeling of the Arizona Game and Fish Department; Bill Block, Joseph Ganey, and Jeff Jenness of the USDA Forest Service's Rocky Mountain Research Station; Carol Chambers and Paul Beier of Northern Arizona University; and Shaula Hedwall of the U.S. Fish and Wildlife Service. We also greatly appreciate the advice and review provided by Barry Noon of Colorado State University, Craig Allen of the U.S. Geological Survey, and Greg Aplet of The Wilderness Society. None of this work could have been attempted, much less completed, without the Ecological Restoration Institute, which provided funding and other forms of support for all phases of this project. Additional funding was received from the USDA Forest Service Joint Fire Sciences Program (a partnership of six federal wildland and fire and research organizations) and the Arizona Game and Fish Department. Special thanks go to Jean Palumbo for copy editing. Obviously, this has been a highly collaborative project, and we appreciate the collective efforts and contributions of all participants, even those whom we have inadvertently omitted here.

LITERATURE CITED

Abrams J (2005a) Report on a participant evaluation of the Western Mogollon Plateau Adaptive Landscape Assessment (WMPALA). Ecological Restoration Inst, Flagstaff.

Abrams J (2005b) Report on a needs assessment for collaborative landscape planning in the White Mountains of Arizona. Ecological Restoration Inst, Flagstaff.

Allen CD, Savage M, Falk DA, Suckling KF, Swetnam TW, Schulke T, Stacey PB, Morgan P, Hoffman M, Klingel JT (2002) Ecological restoration of southwestern ponderosa pine ecosystems: a broad perspective. Ecol Appl 12:1418–1433.

Behan RW (2001) Plundered promise: capitalism, politics and the fate of federal lands. Island Press, Washington.

Cortner HJ (2003) The governance environment: linking science, citizens, and politics. In: Friederici P (ed) Ecological restoration of southwestern ponderosa pine forests. Island Press, Washington, pp 70–80.

Cortner HJ, Moote MA (1999) The politics of ecosystem management. Island Press, Washington.

Covington WW, Moore MM (1994) Southwestern ponderosa forest structure: changes since Euroamerican settlement. J For 92:39–47.

Covington WW, Fulé PZ, Moore MM, Hart SC, Kolb TE, Mast JN, Sackett SS, Wagner MR (1997) Restoring ecosystem health in ponderosa pine forests of the Southwest. J For 95:23–29.

Dahms CW, Geils BW (1997) An assessment of forest ecosystem health in the Southwest. USDA For Serv, Rocky Mountain For Range Exp Stn, Fort Collins. Gen Tech Rep RM-GTR-295.

Dickson BG, Xu Y, Prather JW, Aumack EN, Hampton HM, Sisk TD (2006) Mapping the probability of large fire occurrence in northern Arizona. Landscape Ecol 21 (5):747–761.

Dryzek J S (1990) Discursive democracy: politics, policy, and political science. Cambridge Univ Press, Cambridge, UK.

Durbin K (1999) Tongass: pulp politics and the fight for the Alaska rain forest. Oregon Univ Press, Corvallis.

ERI (2004) Western Mogollon Plateau adaptive management assessment report on initial workshop. Ecological Restoration Inst, Flagstaff. (http://www.forestera.nau.edu/overview_docs.htm)

Finney MA (in preparation) FlamMap. USDA For Serv, Rocky Mountain Res Stn, Fire Sci Lab, Missoula.

Fischer F (2000) Citizens, experts and the environment: the politics of local knowledge. Duke Univ Press, Durham.

Forest ERA and CNF (2005) Coconino National Forest landscape analysis report. Coconino National Forest, Northern Arizona Univ, Flagstaff. (http://www.forestera.nau.edu/updates.htm)

Funtowicz SO, Ravetz JR (1995) Science for the post-normal age. In: Wekstra L, Lemons J (eds) Perspectives on ecological integrity. Kluwer Academic Publishers, Dordrecht.

GCA (2004) Rim country community wildfire protection plan, October 2004. Gila County Authority, Payson Fire Department, Payson, AZ.

GFFP, PFAC (2004) Community wildfire protection plan for Flagstaff and surrounding communities in the Coconino and Kaibab national forests of Coconino County, Arizona. Greater Flagstaff Forest Partnership and Ponderosa Fire Advisory Council, Flagstaff. (http://www.gffp.org)

Glasser H (1995) Towards a descriptive participatory theory of environmental policy analysis and project evaluation. Univ California, Davis. PhD diss.

Hampton HM, Xu Y, Prather JW, Aumack EN, Dickson BG, Howe MM, Sisk TD (2003) Spatial tools for guiding forest restoration and fuel reduction efforts. In: Proc 2003 ESRI Users Conference, San Diego, CA. (http://gis.esri.com/library/userconf/proc03/p0679.pdf)

Hampton HM, Aumack EN, Prather JW, Xu Y, Dickson BG, Sisk TD (2005) Demonstration and test of a spatial decision support system for forest restoration. In: Van Riper C, Mattson DJ (eds) The Colorado Plateau II: biophysical, socioeconomic, and cultural research. Univ Arizona Press, Tucson, pp 47–65.

Harris FW (1995) Science and biodiversity policy. Bioscience 45(6) Suppl:S64-S65.

Holling CS (1978) Adaptive environmental assessment and management. Wiley, New York.

Lee KN (1993) Compass and gyroscope: integrating science and politics for the environment. Island Press, Washington.
Lindblom CE, Woodhouse EJ (1992) The policy making process. 3rd ed. Prentice–Hall, Englewood Cliffs, New Jersey.
Malczewski J (1999) GIS and multicriteria decision analysis. Wiley, New York.
Norton BG (1998) Improving ecological communication: the role of ecologists in environmental policy formation. Ecol Appl 8:350–364.
Pickett STA, White PS (1985) The ecology of natural disturbance and patch dynamics. Academic Press, New York.
Pinchot G (1905) A primer of forestry. Part II—Practical forestry. USDA Bur For, Washington. Bulletin 24.
Porritt J (1994) Translating ecological science into practical policy In: Edwards PJ, May RM, Webb NR (eds) Large-scale ecology and conservation biology. Blackwell Scientific, Oxford, pp 345–353.
Prather JW, Hampton HM, Xu Y, Dickson BG, Dodd NL, Aumack EN, Sisk TD (2005) Modeling the effects of forest restoration treatments on sensitive wildlife taxa: a GIS-based approach. In: Van Riper C, Mattson DJ (eds) The Colorado Plateau II: biophysical, socioeconomic, and cultural research. Univ Arizona Press, Tucson, pp 69–85.
Prather JW, Dodd NL, Dickson BG, Hampton HM, Xu Y, Aumack EN, Sisk TD (2006) Landscape models to predict the influence of forest structure on tassel-eared squirrel populations. J Wildl Manage 70 (in press).
Sarewitz D (1996) Frontiers of illusion: science, technology and the politics of progress. Temple Univ Press, Philadelphia.
Sarewitz D (2004) How science makes environmental controversies worse. Environ Sci Policy 7:385-403.
Schumann ME (2005) ForestERA stakeholder needs assessment, northern New Mexico project, 31 March 2005. (http://www.forestera.nau.edu)
Sisk TD, Hampton HM, Prather JW, Aumack EN, Xu Y, Dickson BG, Loeser MR, Munoz-Erickson T, Palumbo J (2004) Forest Ecological Restoration Analysis (ForestERA) project report 2002–2004. Northern Arizona Univ, Center for Environ Sci and Education, Flagstaff. [Report available in PDF format from: forestera@nau.edu]
Sisk TD, Falk DA, Savage M, McCarthy P (2005) Restoring fire as a keystone process: insights from the pine forests of arid North America. Oral presentation to the 90th Annual Meeting of the Ecological Society of America, 7–12 August 2005, Montreal, Canada.
Sisk TD, Prather JW, Hampton HM, Aumack EN, Xu Y, Dickson BG (2006) Modeling fire and biodiversity to guide ecological restoration of pine forests in arid North America. Landscape Urban Planning 74 (in press).
Swetnam TW, Betancourt JL (1998) Mesoscale disturbance and ecological response to decadal climatic variability in the American Southwest. J Climate 11:3128–3147.
USFWS (1995) Recovery plan for the Mexican spotted owl (*Strix occidentalis lucida*). US Fish Wildlife Serv, Albuquerque.
Walters C (1986) Adaptive management of renewable resources. MacMillan Publishing Company, New York.
Wilkinson CF, Anderson HM (1987) Land and resource planning in the national forests. Island Press, Washington.
Xu Y, Prather JW, Hampton HM, Dickson BG, Sisk TD (2005) Analysis of bias and variance in forest structural mapping using ETM imagery and ground plot data. Poster, 2005 Annual Conference of the American Society for Photogrammetry and Remote Sensing, Baltimore.
Xu Y, Prather JW, Hampton HM, Aumack EN, Sisk TD (2006) Advanced exploratory data analysis for mapping regional canopy cover. Photogramm Eng Remote Sens 72 (1):31–38.

5

Transferring Landscape Ecological Knowledge in a Multipartner Landscape: The Border Lakes Region of Minnesota and Ontario

David E. Lytle, Meredith W. Cornett, and Mary S. Harkness

5.1. Introduction
5.2. Border Lakes Background
    5.2.1. Description of the Border Lakes Landscape
    5.2.2. Ecology of the Border Lakes Landscape
    5.2.3. Land Ownership and Management Goals in the Border Lakes Landscape
5.3. Fundamental Rationales for Working toward a Common Desired Future Condition
    5.3.1. Overlapping Stakeholder Missions
    5.3.2. Scale of Ecological Processes and Land Ownership Patterns
5.4. Programmatic Rationales for Working toward a Common Desired Future Condition
    5.4.1. Agency Requirements to Address a Larger Management Context
    5.4.2. Improved Ability to Leverage Government Funds
5.5. Challenges in Working toward a Common Desired Future Condition
5.6. The Knowledge Transfer Process
    5.6.1. Phase One: Initiate the Collaborative Learning Process
    5.6.2. Phase Two: Pilot Modeling Project

DAVID E. LYTLE • The Nature Conservancy, 6375 Riverside Drive, Suite 50, Dublin, OH 43017, USA   MEREDITH W. CORNETT and MARY S. HARKNESS • The Nature Conservancy, 1101 West River Parkway, Suite 200, Minneapolis, MN 54415, USA.

    5.6.3. Phase Three: Early Review of the Pilot Model
    5.6.4. Phase Four: Build Institutional Support
5.7. Next Steps: Increasing the Scale and Impact of the Partnership
5.8. Lessons Learned
    5.8.1. Summary of Knowledge Transfer Challenges
    5.8.2. Summary of Knowledge Transfer Successes
    Literature Cited

5.1. INTRODUCTION

The Border Lakes landscape of northeastern Minnesota, United States, and northwestern Ontario, Canada, is dominated by a few major, fire-dependent forest ecosystems, and is owned and managed primarily by government agencies with complex hierarchical structures. The Border Lakes Partnership was created to address direct threats to these ecosystems resulting from the severely altered fire regimes in this 2-million-ha, multiple-owner landscape. Following nearly a century of fire suppression, the fire regime of the Border Lakes landscape has been highly altered from its historical range, and the risk of loss of key ecosystem components is high as a result. The fire regime has departed from its historical frequency (an average return interval of 35 to 100 years) to become a regime with multiple return intervals, and dramatic changes in fire size, intensity, severity, and pattern have also occurred (RMRS 1999). Consequently, the plant species composition and the structure of the forest and other ecosystems have shifted substantially. Without the reintroduction of an ecologically appropriate fire regime or a surrogate management practice that emulates that regime, the jack pine-dominated forest ecosystem, a major part of this landscape, may largely disappear from the Border Lakes landscape in the next 50 to 150 years (Heinselman 1973; Paul Tiné, retired, USDA Forest Service, Superior National Forest, personal communication) and others will continue to be highly altered.

Given the human and ecological contexts of the Border Lakes landscape, strategic collaboration among the stakeholders in this landscape will offer many benefits for land management. In order for stakeholders to collaboratively manage land in a particular landscape, both institutional and technical needs must be met. Institutional support for collaboration throughout the agencies is essential both at a local, implementation level and at higher management levels. Sufficient motivation—political will—to work toward common goals must be present and sustained. From a technical perspective, an understanding of the ecology of the ecosystems that dominate a landscape should form the foundation of any collaborative land management effort. In the Border Lakes landscape, the ecological processes that shape the forest ecosystems are a unifying feature of this landscape, and tools for examining the cumulative outcomes of management activities and natural disturbances on forest ecosystems can help members of the Partnership to establish a joint vision for this landscape and identify opportunities for collaboration among stakeholders. The primary goals of this chapter are to illustrate how technical knowledge of and tools for understanding the landscape ecology of this forested region were shared with major public landowners, and to highlight the lessons learned in this knowledge transfer process.

Knowledge transferred in this particular effort included ecological principles and models—specifically, forest ecosystem succession models, an interagency ownership map showing various management objectives, and projections of forest attributes under alternative forest management scenarios derived using the LANDIS software (Mladenoff et al. 1996; Mladenoff and He 1999). A subset of the stakeholders—major public landowners, umbrella groups that coordinate among public agencies and other large private entities, and major nonprofit conservation organizations—were the focus of this knowledge transfer effort. Among this subset of stakeholders, the immediate target audience consisted of natural resource professionals and ecologists within each agency or entity; the longer-term audience for the overall effort included both high-level decisionmakers and on-the-ground implementers. Members of both the immediate and the longer-term target audiences have been involved in the knowledge transfer process described in this case study.

The ultimate goal of the Border Lakes Partnership is for the stakeholders to collaboratively achieve their shared desired future condition for the landscape; the initial knowledge transfer described in this chapter is the first small step in this much larger and longer-term effort. The conceptual ecological models, maps, and projections of forest attributes were identified, developed, and shared within the Border Lakes Partnership to provide a scientific foundation on which the major public landowners and other stakeholders can build a common desired future condition. The spatial and temporal scale of the overarching desired outcome is broad—it will be a long-term effort, and the geographic scale is millions of hectares. Although the overall project is strategic, the Partnership also expects to identify specific implementation steps later in the process. Given the scope of this project and the number and type of stakeholders involved, participants in the Border Lakes Partnership have recognized that developing and beginning to implement a common desired future condition in this landscape will take years; achieving the desired future condition will take decades.

5.2. BORDER LAKES BACKGROUND

5.2.1. Description of the Border Lakes Landscape

The Border Lakes landscape is located in northeastern Minnesota and southwestern Ontario (Figure 5.1). Through a larger regional planning process that included The Nature Conservancy, The Nature Conservancy of Canada, the Ontario Ministry of Natural Resources, and their partners, this 2-million-ha landscape was identified as an important area for conservation because it supports several of the forests and other ecosystems that are representative of the larger Superior Mixed Forest ecoregion (SMFEPT 2002).

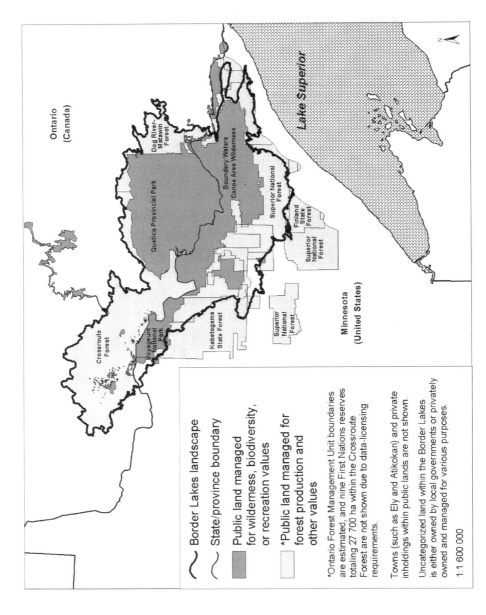

Figure 5.1 Overview of the Border Lakes landscape

5.2.2. Ecology of the Border Lakes Landscape

The Border Lakes landscape is characterized by near-boreal forest ecosystems interspersed with numerous lakes. The dominant potential natural vegetation across most of this landscape is either jack pine–black spruce forest (*Pinus banksiana* Lamb. and *Picea mariana* [Mill.] B.S.P.) or white pine–red pine forest (*Pinus strobus* L. and *Pinus resinosa* Ait.) (White and Host 2000; W. Bakowsky, Ontario Natural Heritage Information Centre and A. Harris, Northern Bioscience Ecological Consulting, personal communication). The mesic aspen–birch–spruce–fir (*Populus tremuloides* Michx., *Betula papyrifera* Marsh., *Picea glauca* [Moench] Voss, and *Abies balsamea* [L.] Mill.) forest ecosystem is the primary potential vegetation in the eastern portion of this landscape, whereas jack pine–aspen–oak (*Quercus* spp.) forest is predominant in northwestern Minnesota.

Fire, wind, and insect outbreaks are the primary disturbances that shape the successional pathways of the predominant forest ecosystems in the Border Lakes landscape. The size, intensity, and frequency of fires historically varied according to the ecosystem type, as well as in response to climatic conditions, fuel loads, topography, and other factors. For example, the average return interval for stand-killing fires in the jack pine–black spruce forest prior to European settlement was approximately 50 to 100 years and ecologically significant fires were relatively large (400 to 4000 ha or more) (Heinselman 1981). Less intense surface fires in the red pine–white pine forests had an average return interval of around 40 years, and ecologically significant fires ranged from approximately 40 to 400 ha. Today, the average fire return interval for these forest ecosystems is significantly longer— approximately 300 to 2000 years across forest types; the average annual area burned is correspondingly much smaller (Heinselman 1981; Ward et al. 2001).

These changes in the fire regimes are a result of fire exclusion and in some areas, a combination of fire exclusion and land cover changes (Frelich 2002; Heinselman 1981; Ward et al. 2001). In contrast, several major blowdown events occurring within the last 30 years were an order of magnitude larger than those documented during the presettlement era (Frelich 2002). Possible reasons for the apparent increase include an increased proportion of older forests, which are more susceptible to windthrow, and more intense storms resulting from global climate change (Frelich 2002). Prior to European settlement, the natural fire and wind regimes created patterns of age structures and species composition that likely limited the extent, intensity, and duration of insect and disease outbreaks. As a result of modern fire suppression, balsam fir is much more abundant and contiguous, allowing spruce budworm (*Choristoneura fumiferana* Clem.) to spread easily over large areas to create intense outbreaks (Heinselman 1973). Researchers and managers have not yet synthesized sufficient quantitative information on the spread, extent, frequency, and duration of historical insect and disease outbreaks in this region; therefore, changes in the average interval between and geographic extent of insect outbreaks have not yet been quantified.

The interactions among fires, wind damage, and insect outbreaks and their spatial and temporal variation have historically created a mosaic of different vegetation growth stages (after Frelich 2002)—characterized by particular combinations of age structure and species composition—in these matrix-forming forests. ("Matrix" or "matrix-forming" refers to ecosystems that dominate a landscape and thus form the matrix within which other smaller scale ecosystems occur. They occur at scales ranging from hundreds of thousands to millions of hectares.) The relative proportion of the different vegetation growth stages has changed dramatically due to a series of historical and ongoing events. Wholesale clearcutting throughout the region from approximately the 1880s to the 1910s created slash loads that subsequently burned in a series of catastrophic, unnaturally extensive and severe fires. Since then, fire has generally been suppressed, and forests have generally been managed for early successional species. With the advent of fire suppression and the management of forests for economic uses, wind and forestry practices now largely determine the mosaic of vegetation growth stages.

Recent land cover classifications in Minnesota and Ontario (MDNR 2002; Spectranalysis Inc. 1999) and other analyses have indicated that early vegetation growth stages of these forest ecosystems are now predominant: a mixture of aspen–birch forest now covers approximately 27% of the entire landscape. Water bodies account for another 25% of this landscape. Late vegetation growth stages of some forest ecosystems, such as the jack pine-dominated forest ecosystem, are also overrepresented at this time. In the matrix-forming jack pine–black spruce ecosystem, conifers historically dominated the species composition (71.5%). However, forest inventory analysis and cooperative stand analysis data indicate that the current overall species composition has shifted to 56% conifers and 44% hardwoods (Brown and White 2002). Without the introduction of an ecologically appropriate fire regime or a suitable surrogate management regime, the jack pine-dominated ecosystems may largely disappear from this landscape within the next 100 years (Heinselman 1973; Paul Tiné, retired, USDA Forest Service, Superior National Forest, personal communication).

5.2.3. Land Ownership and Management Goals in the Border Lakes Landscape

Based on currently available ownership data (BRW Inc. 1999; OMNR 2003a,b, 2004b), approximately 92% of this landscape is owned by public agencies (Figure 5.2). Ownership in the U.S. portion is relatively more complex because of the numerous levels of government (federal, state, county, and municipal) that own parcels of land, and the relative fragmentation of the parcels across ownerships. Federal, state, county, and municipal lands are intermingled, rather than occurring in consolidated parcels. Major public lands on the Minnesota side of the border include the federally owned Superior National Forest, of which the Boundary Waters Canoe Area Wilderness is a part, Voyageurs National Park, and numerous state forests and parks. The vast majority of land in the Border Lakes region of Ontario is Crown (public) land as well, and is composed primarily of Quetico Provincial Park, and

A Multipartner Landscape

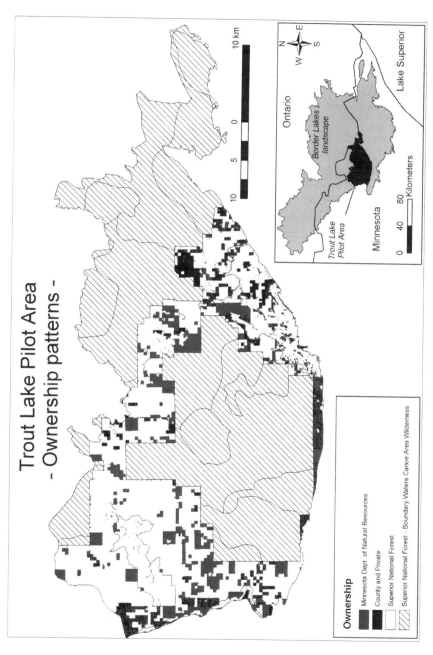

Figure 5.2. Ownership and protected status (data from BRW Inc. 1999; OMNR 2002, 2003a,b, 2004b).

four forest units managed under sustainable forest licenses: the Crossroute, Sapawe, Dog River–Matawin, and Lakehead Forests. Nine First Nations reserves lie within the Crossroute Forest, and First Nations have sovereignty over their land. The pattern of land ownership is relatively simpler in the Ontario portion of the landscape, in part because a single government entity—the province of Ontario—owns roughly 97% of the land and Crown land is broadly divided into fewer, larger units rather than numerous smaller units. As with the First Nations reserves, parcels of privately owned land are embedded within Crown land in this region, but current mapping of the private land is not sufficiently accurate to support precise calculations of its area; however, relative to the amount of Crown land, the total area of private land in the Border Lakes landscape is small. Timber production is a primary management goal on Crown land operated under sustainable forest licenses in Ontario, and in national and state forests in Minnesota. Wilderness areas and parks, such as the Boundary Waters Canoe Area Wilderness and Quetico, are managed primarily for recreation and biodiversity values. Minnesota county and municipal government lands are not highlighted in Figure 5.2, but timber production is a major goal for most of these lands as well. Private lands owned by timber companies are generally also managed for timber production; it is difficult to generalize management goals on other land that is privately owned.

5.3. FUNDAMENTAL RATIONALES FOR WORKING TOWARD A COMMON DESIRED FUTURE CONDITION

The fundamental rationales for the major land-owning agencies, associated coordinating bodies, and nonprofit conservation groups in this region to work together toward a common desired future condition are twofold: (1) their missions significantly overlap or relate to each other, and they generally recognize the importance of all the current uses of this forested landscape, and (2) the driving ecological processes that shape the landscape's forest ecosystems operate at scales that will frequently cross one or more ownership boundaries. Under such circumstances, it is at best inefficient and at worst counterproductive when these stakeholders fail to work together (e.g., Kutas et al. 2002).

5.3.1. Overlapping Stakeholder Missions

Stakeholders in the Border Lakes landscape include numerous public and private entities that own or manage large parcels of land in this region, as well as organized groups with an interest in how the land is managed, individual residents of the Border Lakes region, and the broader community (such as citizens of Minnesota and Ontario) who are interested in the landscape and its various uses. The subset of stakeholders that are the focus of the initial knowledge transfer project described in this chapter are summarized in Table 5.1; this is not an exhaustive list of stakeholders, and focuses only on areas of greatest mission overlap.

5.3.2. Scale of Ecological Processes and Land Ownership Patterns

Fires, windstorms, and insect outbreaks in this landscape vary greatly in size, ranging from hundreds to hundreds of thousands of hectares. The size of contiguous ownership parcels is also variable. Quetico Provincial Park, at nearly 476 000 ha, is the largest single contiguous ownership parcel in this landscape. However, many of Minnesota's state forests and parks, some of Ontario's Sustainable Forest License areas, Voyageurs National Park, and smaller units of the Superior National Forest and the Boundary Waters Canoe Area Wilderness range from roughly 10 000 to 75 000 ha in size. The area affected by a large disturbance event will almost never fit neatly within even the largest contiguous ownership parcels, and will typically cross many smaller ownerships. For example, the 4 July 1999 blowdown affected 237 000 ha of this landscape; 150 000 ha of this affected area were within the Boundary Waters Canoe Area Wilderness, but the remainder were spread across parts of the Superior National Forest, Quetico Provincial Park, and other Crown lands in Ontario (USDA Forest Service 2001). Average- or smaller-sized disturbances will also often cross ownership boundaries in this landscape, as there are numerous smaller and midsize parcels (see Figure 3 in Heinselman 1973 for maps showing fires that extended beyond the Boundary Waters Canoe Area Wilderness).

Managing individual parcels strictly within ownership boundaries is inconsistent with the current understanding of the scale and pattern of the driving ecological processes in this landscape. Instead of viewing the interaction of management practices and disturbance events within the smaller context of a single ownership parcel, it can be helpful to consider those interactions at the scale at which they occur—which means including adjacent ownerships in such considerations. Given the varied scales of disturbance events and of ownership patterns, and the shared goals in shared ecosystems, it makes sense for adjacent landowners to coordinate their management goals, management activities, and responses to and anticipation of major disturbance events.

5.4. PROGRAMMATIC RATIONALES FOR WORKING TOWARD A COMMON DESIRED FUTURE CONDITION

Practical, specific rationales for collaborating include the internal requirements of public agencies to account for the larger context of their management activities, and an improved ability to leverage funds from various government sources that place a high priority on collaborative projects.

5.4.1. Agency Requirements to Address a Larger Management Context

Many of the public agencies that own land in the Border Lakes landscape have broader mandates to address landscape-scale contexts. For example, the U.S. National Environmental Policy Act requires Superior National Forest (and all other national forests) to consider the landscape context in their forest management plans,

Table 5.1. Stakeholder missions related to recreation, biodiversity, timber production, and forest protection (the prevention and suppression of excessive damage to forest resources caused by fire, insects, and diseases). Primary missions are represented by dark gray, secondary missions by light gray

Agency	Mission priority				Mission statement
	Recreation	Biodiversity conservation	Timber production	Forest protection	
Major land managers[a]					
USDA Forest Service: Boundary Waters Canoe Area Wilderness	■ dark	■ dark		■ dark	The Wilderness Act established the National Wilderness Preservation System in order to "...secure for the American people of present and future generations the benefits of an enduring resource of wilderness".
USDA Forest Service: Superior National Forest	▨ light	▨ light	■ dark	■ dark	"The mission of the USDA Forest Service is to sustain the health, diversity, and productivity of the Nation's forests and grasslands to meet the needs of present and future generations."
U.S. National Park Service: Voyageurs National Park	■ dark	■ dark		■ dark	"The National Park Service preserves unimpaired the natural and cultural resources and values of the national park system for the enjoyment, education, and inspiration of this and future generations. The Park Service cooperates with partners to extend the benefits of natural and cultural resource conservation and outdoor recreation throughout this country and the world."
Minnesota Department of Natural Resources	■ dark	■ dark	■ dark	■ dark	"Our mission is to work with citizens to conserve and manage the state's natural resources, to provide outdoor recreation opportunities, and to provide for commercial uses of natural resources in a way that creates a sustainable quality of life."
Minnesota Department of Natural Resources: Division of Forestry (state forests)	▨ light	▨ light	■ dark	■ dark	"Through shared information, technology, and understanding, we empower others and ourselves to sustain and enhance functioning forest ecosystems; provide a sustainable supply of forest resources to meet human needs (e.g., material, economic, and social); protect lives and property from wildfires; and provide a dollar return to the permanent school trust."
Minnesota Department of Natural Resources: Parks and Recreation (state parks)	■ dark	■ dark		■ dark	Parks and Recreation "will work with people to provide a state park system which preserves and manages Minnesota's natural, scenic, and cultural resources for present and future generations while providing appropriate recreational and educational opportunities."

A Multipartner Landscape

Organization	Mission/Vision
Ontario Ministry of Natural Resources: Parks and Recreation Quetico Provincial Park	"To ensure that Ontario's provincial parks protect significant natural, cultural, and recreational environments, while providing ample opportunities for visitors to participate in recreational activities." As a wilderness park, Quetico's specific mission is to "preserve Quetico Provincial Park, which contains an environment of geological, biological, cultural and recreational significance, in perpetuity for the people of Ontario as an area of wilderness that is not adversely affected by human activities."
Ontario Ministry of Natural Resources: Forestry	"To ensure excellence in the management and protection of Ontario's forests and the provision of specialty resource management services." Vision: "Sustainable Forests—healthy forests providing balanced environmental, social and economic benefits now and forever."
Other stakeholders	
The Nature Conservancy	"To preserve the plants, animals and natural communities that represent the diversity of life on Earth by protecting the lands and waters they need to survive."
The Nature Conservancy of Canada	"NCC protects plants, animals and natural communities by safeguarding the land and waters they need to survive."
Minnesota Forest Resources Council	Vision: "Minnesota forests are managed with primary consideration given to long-term ecosystem integrity and sustaining healthy economies and human communities. Forest resource policy and management decisions are based on credible science, community values, and broad-based citizen involvement. The public understands and appreciates Minnesota's forest resources and is involved in and supports decisions regarding their use, management and protection."
Minnesota Incident Command System, Prescribed Fire Working Team/Minnesota Interagency Fire Center	The team and center comprise U.S. federal and state agencies, including the Minnesota Department of Natural Resources, USDA Forest Service, National Park Service, U.S. Fish and Wildlife Service, and Minnesota Department of Public Safety, and provide coordination, education, and implementation of the Incident Command System to support responses to fires and all risk incidents in Minnesota and nationally.

[a] An agency or other entity that owns and manages more than 20,000 ha in the Border Lakes landscape.

timber sales, wildlife improvement plans, and other activities. The same will be true for Voyageurs National Park (and all national parks) when they update their management plans. In addition, the public agencies and private industry in Minnesota have made commitments to assess the landscape-level impacts of their forest management practices. The Minnesota Department of Natural Resources, for example, has recently decided to attain certification under both the Forest Stewardship Council and the Sustainable Forest Initiative for its forests and forestry operations; certification by either program explicitly requires a consideration of the landscape context in their forest management decisions. The Ontario Ministry of Natural Resources' Crown forest planning guidelines (OMNR 2004a) require that "management objectives in each forest management plan be compatible with the sustainability of the Crown forest," and the Crown forests surrounding Quetico Provincial Park are required to consider the impacts of their management on Quetico.

Agency requirements for considering the landscape context in their management activities appear to be part of a longer-term trend within many government agencies to strive for greater efficiency and effectiveness through collaboration. In the United States, the 2001 Federal Wildland Fire Policy, which will be implemented through the National Fire Plan, recognizes that "Federal, State, tribal, local, interagency, and international coordination and cooperation are essential" and subsequently requires that "Fire management planning, preparedness, prevention, suppression, fire use, restoration and rehabilitation, monitoring, research, and education will be conducted on an interagency basis with the involvement of cooperators and partners." In the Border Lakes region, additional collaborative efforts are underway to address slightly different sets of issues at the landscape scale. In Ontario, a formal network of researchers, agencies, and organizations is conducting a variety of research projects relating to sustainable forest management within the "Legacy Forest," which includes the southern half of the Dog River–Matawin Forest and the adjacent Quetico Park; one of the major long-term goals of the network is to understand the landscape-level impacts of various forest management practices (http://www.legacyforest.ca/). Minnesota's Forest Resources Council continues to coordinate regional forest management planning across the state. In addition, effective coordination of fire management is taking place among state, provincial, and national public agencies in northern Minnesota and northwestern Ontario, partly as a result of the 1999 blowdown, which greatly increased the future risk of large fires. Most of the partners involved in the present Border Lakes effort have been involved in the aforementioned efforts, and some are also part of Ontario's Legacy Forest project; these previously established initiatives have created a solid precedent for collaborating on issues that transcend ownership boundaries.

5.4.2. Improved Ability to Leverage Government Funds

Many government grants set a high priority or even a requirement for funding proposals to include appropriate, effective partnerships and collaborations rather than directing funds to individual entities working independently. The United States

National Fire Plan funding guidelines (www.fireplan.gov) specify that an important criterion for a project to successfully compete for funding is having large-scale, broad partnerships with clear local community support. It is not uncommon for successful National Fire Plan proposals to list six or more "secondary partners" in addition to the recipient of the funding award. In 2003, US$426 million of funding from the Healthy Forest Restoration Act (P.L. 108-148; http://www.healthyforests.gov/initiative/legislation.html) was used to implement projects for the reduction of hazardous fuels across the nation. This Act emphasizes improved coordination among agencies, such that federal fire activities, including rehabilitation and restoration of land, are integrated with those of states, First Nations, and local governments (www.fireplan.gov/healthyforest/index.html). Canada's Sustainable Forest Management Network funds projects that focus on questions such as how to develop tools for scenario planning and assessment under different combinations of multiple-use forest values and land-use intensities. It emphasizes innovative proposals involving interdisciplinary research teams that assess forest management strategies and alternatives using science-based criteria (http://sfm-1.biology.ualberta.ca/english/research/en_cfp.htm). Given this trend, stakeholders who can demonstrate truly cooperative projects in a landscape context have a better chance to obtain funding from a wider variety of sources.

5.5. CHALLENGES IN WORKING TOWARD A COMMON DESIRED FUTURE CONDITION

Many of the specific challenges to achieving a common desired future condition in the Border Lakes landscape stem from two inherent political characteristics of this landscape: there are multiple political jurisdictions (Canada and the United States, Ontario and Minnesota), which include an international border, and numerous public and private landowners, which include multiple levels of government agencies (national, state and provincial, and local). In general, the more stakeholders and landowners involved, the more difficult it is to plan and implement landscape-scale conservation and management activities (see Kutas et al. 2002 and Pedynowski 2003).

In the Border Lakes landscape, many of the stakeholders are large government agencies, and although their missions overlap significantly, each stakeholder must nonetheless accomplish its particular mission and answer to its local, state or provincial, or national constituency. In addition, it is often challenging for large organizations to respond quickly to new ideas, tools, or approaches. With so many geographically scattered stakeholders in a large landscape, the logistical difficulties and the costs of coordination and of face-to-face collaboration are greatly magnified. Bureaucratic constraints on staff travel across state, provincial, or international borders can further impede active collaboration.

Although public agencies that own and manage land have internal mandates to consider in their management in the context of the larger landscape, to date they have generally lacked the tools and the human and financial resources to do so in a coordinated, strategic way. As mentioned earlier in this chapter, many of the agencies that own land in the Border Lakes landscape have already conducted independent planning efforts to address their institutional missions. Although some of the plans addressed the landscape context of their proposed actions by assessing the actions of nearby landowners, these assessments were limited because they only evaluated the general trends in forest management by neighboring land management agencies; detailed assessments of the spatial and temporal patterns of management activities typically are not conducted. In addition, agency staff are often so consumed with their agency's planning efforts, in addition to their regular duties, that they have very few opportunities to address cross-boundary, landscape-level issues. There is thus an incomplete picture of how management plans add up to a cumulative, landscape-level whole.

The number of and variation in stakeholders also means that the consistent data sets necessary to build the scientific foundations for this kind of collaboration may be lacking, or that it may require substantial effort to develop baseline consistency among existing data sets. In some cases, attempting to create such consistency among data sets will oversimplify them to the point where they are no longer useful for analysis. Truly consistent data sets may not exist, especially when working across major political boundaries (international, state, or provincial). The quality of the data often varies across jurisdictions—even if the data are consistent, their quality (e.g., accuracy, level of detail) may be insufficient. When high-quality data do exist, they may not be readily available because of their format, proprietary nature, or differences in data-sharing customs.

The third challenge is somewhat independent of jurisdictions and ownership. Determining the most effective management strategies for moving a landscape from the current to the desired condition requires a degree of tolerance of scientific uncertainty and modeling skills that many practitioners lack. Nonetheless, the scientific uncertainty associated with modeling principles and tools presented much less of a barrier than the larger issue posed by the number of jurisdictions and stakeholders.

Finally, it is important to recognize that beyond the challenges identified above, fundamental challenges are also posed by the overall political and social context of the landscape. Moving this landscape to a mutually agreeable desired future condition will ultimately require the support of not only the major landowners, but also of local communities, smaller landowners, and other interested parties.

5.6. THE KNOWLEDGE TRANSFER PROCESS

A cooperative project between The Nature Conservancy, the USDA Forest Service, and the Department of the Interior led to the creation of the Fire Learning Network in 2002 in an effort to overcome implementation barriers to ecologically appropriate projects that reduce hazardous fuels and restore fire-dependent ecosystems (http://tncfire.org/training_fire.htm). A series of collaborative forums organized by the Network provided a framework and approach for knowledge transfer in the Border Lakes Partnership. The importance of a collaborative learning approach to knowledge transfer is illustrated later in this section by descriptions of each phase of the project. Detailed descriptions include the challenges encountered with each step and how those challenges were addressed. Key steps are summarized in Table 5.2. Although considerable progress has been made during the last 2 years, the first four phases represent a collaborative learning process that we hope will provide the foundation for a project that spans multiple years. The overarching goal is to engage partners in developing a long-term, large-scale vision that they will implement through the identification of strategic, collaborative opportunities.

5.6.1. Phase One: Initiate the Collaborative Learning Process

We initiated the collaborative learning process in the winter of 2003 with a knowledge transfer forum held under the auspices of the Fire Learning Network. Participants included representatives from the USDA Forest Service (Superior National Forest and North Central Research Station), the National Park Service (Voyageurs National Park), Ontario Ministry of Natural Resources (Quetico Provincial Park), the Minnesota Department of Natural Resources, and The Nature

Table 5.2. Timing of the key steps in the knowledge transfer process for the first two years of the Border Lakes Partnership

Knowledge transfer process	Time frame	Steps
Phase one: initiate the collaborative learning process	Winter 2003	Hold the first knowledge transfer forum, with an emphasis on ecological models and qualitative desired future condition.
Phase two: develop a pilot model	Spring and summer 2003	Assemble data layers from partner agencies and gather partner input on assumptions for the pilot modeling exercise.
	Fall 2003	Hold a second knowledge transfer forum, with an emphasis on developing a spatially explicit desired future condition.
Phase three: early review of the pilot model	Winter 2004	Present the results of the pilot modeling exercise to each of the four key agency partners for input and refinement.
Phase four: build institutional support	Spring 2004	Begin building institutional support for expanding the collaborative modeling exercise to the whole landscape, including meetings with agency technical leadership and the development of communication products such as a fact sheet.
	Summer 2004	Hold a third knowledge transfer forum, with an emphasis on refining a spatially explicit desired future condition and developing strategies for overcoming challenges to future implementation.
Next steps: increase the scale and impact of the initiative	Winter and spring 2005	Assemble data layers for expanding the modeling to encompass the full 2-million-ha landscape.
	Winter and spring 2005	Hold a meeting of technical team partners to discuss future uses of scenario modeling, including the selection of a desired future condition.
	Fall and winter 2005	Hold a fourth knowledge transfer forum, with an emphasis on selecting a collaborative desired future condition and examining options for collaborative, cross-boundary projects.

Conservancy. Prior to the forum, we sought participant input on the forum's goals, the conceptual ecological models to be considered, and a qualitative description of the desired future condition. We also engaged the Minnesota Incident Command System, the lead interagency fire organization in the state. The early approval of the System's Prescribed Fire Working Team was crucial, and they assisted with publicizing the forum. As a result, agency attendance was excellent, even though there was some confusion about the purpose of the forum. Several conceptual ecological models formed the basis for the Border Lakes discussion with the goal of establishing a common language about ecological processes and successional pathways (e.g., Figure 5.3; Brown and White 2002). Although opinions varied on the broad application of the models, the models nonetheless provided an important conceptual starting point for developing a common understanding of natural disturbance regimes in the Border Lakes landscape.

How to develop a long-term (100 years), quantitative set of spatially explicit description of the desired future conditions for the landscape was a more difficult topic for the group to tackle. Although in principle all partners acknowledged the advantages of working collaboratively, they focused initially on programmatic benefits such as securing federal monies for cross-boundary projects. Before committing to collaboration on a common vision, partners requested that The Nature Conservancy facilitate the modeling of several alternative forest management scenarios across the Border Lakes landscape to inform their development of a desired future condition. They further recommended that we start by modeling forest management scenarios for a smaller, pilot project area to test the applicability of this modeling approach for the entire landscape. To move the modeling process forward, a small team of the partners volunteered to identify a suitable landscape for the pilot project, conduct the pilot assessment, and report back to the full Border Lakes Partnership on their progress and the initial results.

The high degree to which partners were engaged in the discussion emerged as the primary achievement during Phase One of the project. However, the discussion tended to leap ahead into strategy development rather than laying the groundwork for establishing a joint vision. Our experience is that this is a common tendency among land managers, particularly for a large landscape with complex ownership patterns. Rather than forcing the visioning process, we documented strategy suggestions throughout Phase One to assist in later phases of the collaboration.

5.6.2. Phase Two: Pilot Modeling Project

Subsequent to the decision to proceed with a modeling approach, the modeling team identified a number of "sideboards." First, they agreed that the pilot landscape should contain a sufficient range of initial conditions (e.g., forest structures, stand compositions, and stand ages) and management objectives to ensure that the model's projections would provide insights into the entire Border Lakes landscape. Second, they agreed that the modeling exercise should project the future of several management scenarios that reflected different sets of assumptions about the management activities undertaken within the pilot project area. The resulting landscape projections could

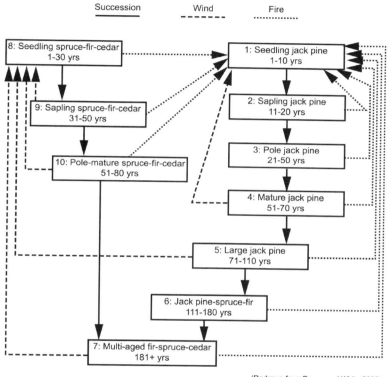

Figure 5.3. Example of the conceptual state and transition model used for the knowledge transfer exercise, with natural resource managers as the intended audience. Each box describes a unique vegetation growth stage (numbered 1 through 10), with the stand age determined by time since last disturbance. Arrows show changes in forest age, composition, and structure resulting from succession and canopy-replacing wind and fire disturbances.

thus provide a range of outcomes for the Border Lakes partners to use in the development of the desired future condition for the landscape. These management scenarios were to be based on the planning work of the partner management agencies, although the modeling team also felt strongly that the management scenarios should not be constrained by these forest management plans if there was reason to model the use of alternative management tools and techniques.

The modeling team selected the Trout Lake land type association in Minnesota (hereafter referred to as the Trout Lake pilot area; Figure 5.4) as its pilot landscape. In this context, a land type association is a fine-scale ecological map unit within the ecological land classification system (Bailey 1995) developed by the USDA Forest

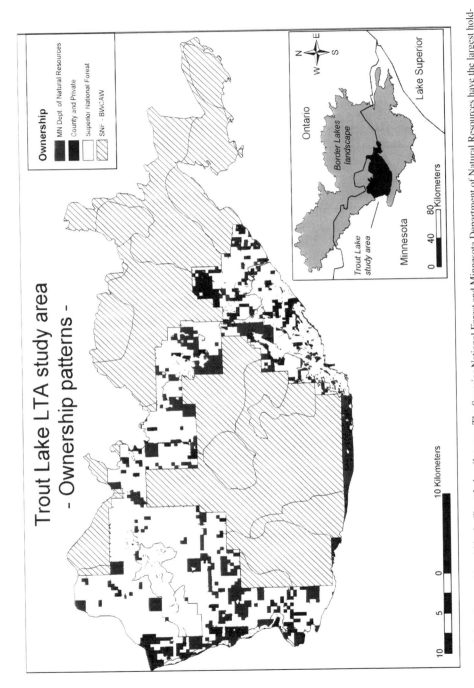

Figure 5.4. Land ownership within the Trout Lake pilot area. The Superior National Forest and Minnesota Department of Natural Resources have the largest holdings. The "Other" category includes county and private nonindustrial ownerships.

A Multipartner Landscape

Service. This 160 000-ha area is centrally located within the Border Lakes landscape, and is similar to the larger landscape in that most of the area is publicly owned and is managed for multiple objectives ranging from wilderness to intensive timber production. The ownership pattern is fragmented, with a mixture of lands managed by the Minnesota Department of Natural Resources and the Superior National Forest dominating the landscape. The selection of a single element within the ecological classification system also worked well as a basis for the simulation model because it simplified the development of site-specific model parameters.

The team used the LANDIS forest dynamics simulation software (Mladenoff et al. 1996; Mladenoff and He 1999) to carry out the landscape simulations requested by the Border Lakes partners. This tool simulates succession, fire, harvesting, and wind damage over large areas (10^4 to 10^6 ha) and long time spans (10 to 1000 years), and is thus well suited for the type of analysis the Border Lakes partners desired. An overview of the model is presented in Table 5.3. The model allowed partners to

Table 5.3. An overview of the attributes and data requirements of the LANDIS software

Parameter	Specifics
Purpose	A spatially explicit simulation model of landscape-level forest dynamics
Spatial domain	10 000 to >1 000 000 ha
Temporal domain	100 to >1000 years, in 10-year time steps
Processes simulated	Seed dispersal
	Competition
	Fire
	Wind
	Harvesting
	Biological disturbances such as disease and insects
	Fuel accumulation and decay
	Biomass accumulation and decay
Required data	Maps of forest composition and age
	A map of land types
	For fire: return intervals and mean, minimum, and maximum fire sizes
	For wind: return intervals and mean, minimum, and maximum areas of storm damage
	For harvesting: minimum harvestable area, species preferences, adjacency rules, reentry periods, and many other relevant parameters
	Species characteristics
Model output	Species composition and age
	Forest stand maps
	Fire maps
	Wind maps
	Harvesting maps
	Disease and insect outbreak maps
	Fuel maps
	Biomass maps
Model structure	Raster-based, with variable cell size (10 to 100 m)
	Within each cell, species cohorts are tracked by age
Developers	David Mladenoff (University of Wisconsin–Madison)
	Hong He (University of Missouri–Columbia)

develop spatially and temporally explicit examinations of the goals of the partner agencies, and of the important interacting processes (timber harvesting, forest succession, and fire) that affect the Border Lakes landscape. Reinforcing the need for a spatial analysis was the recognition that some land management objectives are incompatible when arranged in certain spatial patterns (e.g., two adjacent parcels, one managed for a wilderness objective and the other for intensive timber production), and that a nonspatial assessment could not identify such incompatibilities. However, use of the LANDIS model created new challenges for knowledge transfer. Parameterization and implementation of the modeling scenarios in LANDIS requires expert modeling skills that take months for even a highly trained individual to learn. As such, it was not realistic to expect the Border Lakes partners to take this tool and run their own scenarios. Thus, we were challenged to simplify this complexity sufficiently that partners could remain active participants in the use of this modeling tool and that key stages of the modeling were as transparent as possible. At this stage, a core group of two staff from The Nature Conservancy and the USDA Forest Service's North Central Research Station took responsibility for moving the project forward. The Nature Conservancy focused on cultivating relationships with the primary partners, while the North Central Research Station took the technical lead in data acquisition and modeling.

The management scenarios selected by the modeling team included components of each of three core elements: the management strategy, the role of prescribed fire, and the role of wildfire. The management strategy contrasted the current management plans of the partner agencies with newly proposed (but not yet implemented) management plans. In 2004, the Superior National Forest finalized and implemented its revised forest plan (USDA Forest Service 2004). The model scenarios described here are, however, based on the Draft Forest Plan Revision (i.e., the best information available at the time of the modeling exercise; USDA Forest Service 2004). For clarity, the Draft Forest Plan Revision is referred to henceforth as the "proposed plan," and the now-outdated Superior National Forest Plan of 1986 (USDA Forest Service 1986) is referred to as the "current plan." Within the Trout Lake pilot area, the current and proposed plans differed substantially in their objectives (Figure 5.5a,b).

The second and third elements contrasted the effects of the current policy of fire suppression with alternative strategies in which prescribed fire and managed wildfire were used to achieve management goals. These elements reflected the modeling team's concerns about the 80-year history of fire suppression within the Border Lakes landscape. Of the eight possible combinations of options, three management scenarios were selected for analysis: the current management strategy combined with fire suppression (i.e., the status quo), the proposed management strategy combined with fire suppression (i.e., the strategy likely to be implemented on the ground in the near future), and the proposed management strategy combined with the use of both prescribed fire and managed wildfire. In addition to these three management scenarios, a fourth scenario paired a strategy of no timber harvesting with fire suppression. This scenario served as a control and was added to assess the impact of a

A Multipartner Landscape

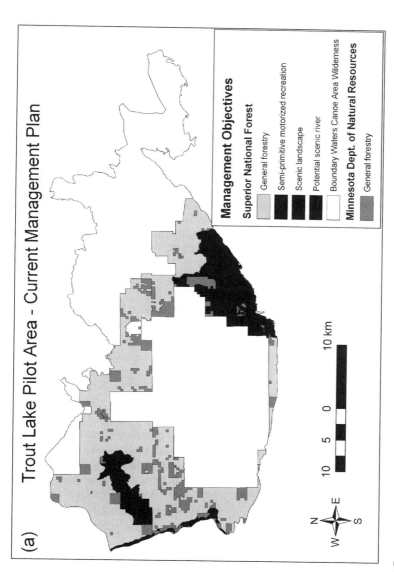

Figure 5.5. (a) Current management objectives within the Trout Lake pilot area. The "General forestry" objective is achieved through the use of a variety of silvicultural techniques, but relies most heavily on even-aged management practices, including clearcut harvesting. The "Scenic landscape," "Potential scenic river," and "Semi-primitive motorized recreation" objectives balance timber production with recreational goals, and make greater use of partial harvesting and uneven-aged management practices. No timber harvesting is allowed within the Boundary Waters Canoe Area Wilderness. Fire suppression is permitted to support all management objectives.

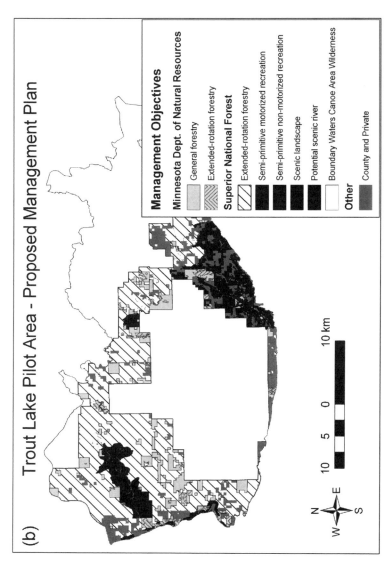

Figure 5.5. Cont'd (b) Management objectives proposed within the Superior National Forest's draft forest management plan, and the Minnesota Department of Natural Resources draft Border Lakes subsection forest management plan. The "Extended-rotation forestry" objective utilized longer timber harvest rotations than the "General forestry" objective, and also relied less heavily on even-aged management. The "Semi-primitive motorized recreation," "Semi-primitive non-motorized recreation," "Scenic landscape," and "Potential scenic river" objectives balance timber production with recreational goals, and make greater use of partial harvesting and uneven-aged management practices. The "General forestry" objectives are the same as described in (a).

"hands-off" management approach, which provided a useful reference point for discussions between the management partners and external stakeholders concerned with the effects of forest management.

The initial data for the LANDIS modeling project (including forest composition, structure, and age; maps of management objectives; and descriptions of the techniques used to meet the objectives) were provided to the modeling team by the Border Lakes partners. These data are collected and maintained in a digital format by the largest public and private forest management organizations, and thus required relatively little manipulation before they could be fed into the model. Data on harvesting rates and techniques are somewhat more difficult to convert into model-accessible formats, but even this process was relatively easy because the members of the modeling team drew on the expertise within their own agencies to provide the necessary data. The parameterization phase of the modeling effort also proved helpful to the modeling team for a completely different reason: it allowed those unfamiliar with LANDIS to learn more about its structure, assumptions, and outputs. In turn, this knowledge helped the team define the types of data to be produced by the project, and the types of conclusions these data would support.

Once the parameterization was complete and the team had determined the types of outputs the model would be used to generate, the simulations were conducted and the raw data were compiled and presented to the modeling team for evaluation. Although the original purpose of the evaluation was to refine the outputs that would be presented at a meeting of the Border Lakes partners, it also served a second, and unintended, purpose: it allowed the modeling team members to compare the model projections with their own expectations. The model projections were uniformly seen as reasonable and logical outcomes of the four management scenarios, and thus were essentially "validated" by the modeling team. As a result, the team members were willing to promote the Border Lakes Partnership process within their own organizations, to arrange access to key decisionmakers, and to provide funding to continue the project. In retrospect, the opportunity to validate model projections was a critical step in the progress of the Border Lakes Partnership.

Model projections provided insights into the potential effects of forest management within the Trout Lake pilot area, and highlighted the importance of a landscapewide assessment in measuring these effects. Among the most interesting results from the Trout Lake modeling process are those that place the ability to achieve the goals of the management agencies (as outlined within the agency planning documents) into a broader landscape context. For example, one of the goals contained within the management scenario based on the proposed management plans was an increase in the area occupied by red pine and white pine. This goal was shared by both the Superior National Forest and the Minnesota Department of Natural Resources, and their planning documents show that where they actively manage the landscape (i.e., the areas outside the Boundary Waters Canoe Area Wilderness) through timber harvesting and other tools, this goal is being met (MDNR 2004; USDA Forest Service 2004). However, when the unmanaged portions of the Trout Lake pilot area were also included in the analysis, there was no evidence of an increase in red pine and white pine stands at the landscape scale, and little difference between the current and proposed management plans (Figure 5.6a). At the landscape scale, therefore, the maintenance of existing red pine and white pine stands and the creation of new ones through active management was balanced by the loss of stands due to natural successional processes in unmanaged portions of the landscape. The goal defined for the actively managed portions of the landscape was thus not achieved when considered in the larger landscape context. However, using prescribed burns and wildfire as a management tool could increase the abundance of both pines across the landscape, both when stands of these species are considered and when the abundance of both species across all stand types is considered (Figure 5.6a,b).

Figure 5.7 provides a second example of the importance of the landscape context in developing the desired future condition by depicting a small portion of the pilot area under the proposed management plans of the Minnesota Department of Natural Resources and the Superior National Forest. As the figure demonstrates, management objectives can vary greatly across even a small area, and potentially conflicting management objectives may exist in close proximity. The juxtaposition of wilderness, general timber production, extended-rotation forestry, and recreational objectives within a small area may create challenges for land managers attempting to achieve goals related to wildlife, aesthetics, and recreation, not because of their own management actions, but because of management actions taken in adjacent land ownerships.

The examples of red pine and white pine stands and of conflicting management objectives in adjacent ownership parcels are not intended as a criticism of the planning processes of the Border Lakes project partners. Indeed, within the context of their planning activities, the agencies with ownership in the Trout Lake pilot area were successful in meeting their goals. Rather, these examples are intended to highlight two points: (1) understanding the context within which plans are developed and implemented is critical, as the ability to meet goals changes when the spatial or temporal scale of the analysis changes, and (2) a landscape-level analysis can add value to the planning processes of individual stakeholders by changing the scale, and thus the perspective, of the analysis. The experience of the Trout Lake modeling group suggests that these points were lost neither on the Border Lakes partners nor on key decisionmakers within the partner agencies. For example, when Figure 5.7 was shown to partners and decisionmakers, there was a ready acknowledgment of potential management incompatibilities and, more significantly, a willingness to consider strategies that would increase coordination of management actions across ownership boundaries and possibly to spatially rearrange the management objectives in order to reduce management conflicts.

5.6.3. Phase Three: Early Review of the Pilot Model

Flexibility in meeting with partners individually and in small groups was especially important during the review of the pilot model, in addition to during the formal forums held by the Fire Learning Network. During Phase Three, two distinct audiences for

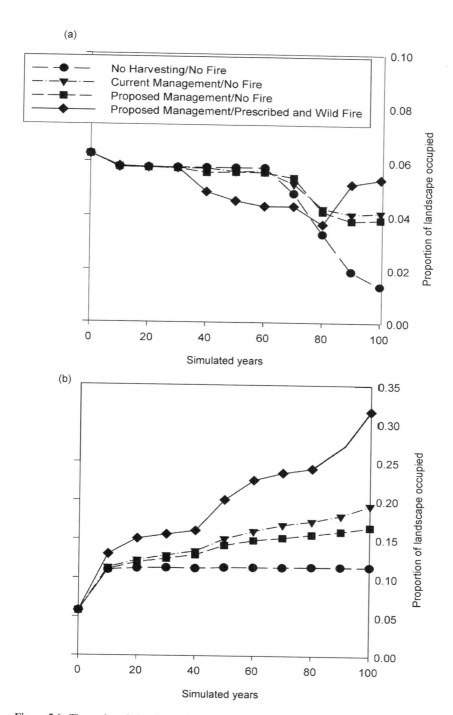

Figure 5.6. The projected abundance of red pine and white pine across the Trout Lake pilot area, as a proportion of the landscape, under four management scenarios: (a) red and white pine stands and (b) red and white pine as component of all stand types.

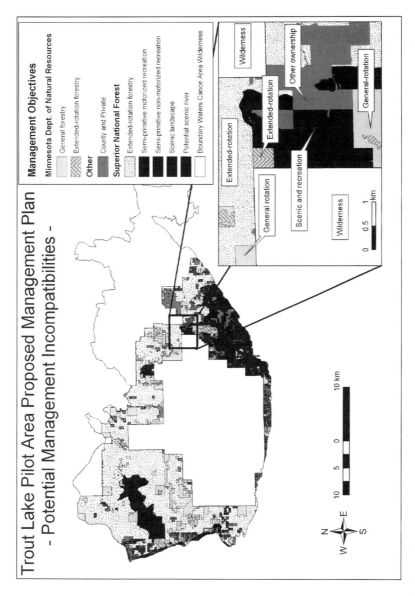

Figure 5.7. Potentially incompatible land management objectives under the proposed management scenario. The inset highlights part of the Trout Lake pilot area to show the juxtapositioning of land management objectives.

knowledge transfer and communication emerged: natural resource managers and decisionmakers. Flexibility in the size and timing of update meetings was essential in order to obtain feedback from resource managers. Agency partners, including the Superior National Forest, the Minnesota Department of Natural Resources, and Voyageurs National Park, critiqued the pilot model results during two separate sessions. Representatives from the Department of Natural Resources and Quetico Provincial Park were unable to attend either session, necessitating individual updates for their staff. Results from the pilot project were well received, with several insights gained into the ways in which the approach could be helpful to each agency's mission, planning efforts, and land management. We considered this to be successful knowledge transfer, and a breakthrough in the project. The pilot project enabled us to establish internal advocates and identify early implementers among the technical partners.

Although well-received by the Voyageurs National Park staff, the pilot project was less directly relevant to their routine management and planning because no National Park lands were located within the pilot area. Moreover, the park's recent staff changes required the team to update them on the Border Lakes project's background and history. We expect that expansion of the model to cover the entire landscape will create opportunities for greater participation on the part of park staff in the future.

5.6.4. Phase Four: Build Institutional Support

Building on the renewed enthusiasm among the resource managers permitted additional outreach to institutional decisionmakers for continued, landscape-level modeling. We developed a fact sheet to explain the pilot modeling results and the potential for the information to inform existing management plans and future collaborative activities (TNC 2004). With the cooperation of a key Superior National Forest technical team member, outreach began with National Forest decisionmakers and the Minnesota Forest Resources Council. This Council, the primary large-scale goal-setting body for forests in Minnesota, encompasses government agencies, nongovernmental organizations, and timber-industry partners. The Council invited core team members to present their preliminary results to a broader audience, leading to a grant awarded to the team, matched by the National Forest, to continue the ecological modeling work. Again, we considered this to be a successful knowledge transfer, made easier by the existing technical backgrounds of National Forest and Council leadership.

Building institutional support was more difficult than expected within The Nature Conservancy as a result of communication and priority-setting challenges. Although the 3-year strategic plan of the Minnesota chapter of The Nature Conservancy identified the Border Lakes landscape as a priority, initial participation on the part of local program staff was limited, largely due to a lack of staff. Communication among key staff members also needed improvement, and the definition of roles and expectations needed to be clarified. Issues surrounding communication were resolved in part by creating a vision document for The Nature Conservancy's role in fire issues related to northern forests. To advance the project, it was also necessary to distinguish clearly between two main project objectives: developing ecological models to create a scientific underpinning for modeling and identifying appropriate strategies for implementation. A long-term goal of greater partner involvement and leadership in building institutional support within participating agencies was also developed.

5.7. NEXT STEPS: INCREASING THE SCALE AND IMPACT OF THE PARTNERSHIP

Despite the challenges encountered during the first 2 years of the Border Lakes Partnership, scenario modeling for the Border Lakes landscape shows great promise for developing a collaborative vision. Based on the response from the project partners, we plan to take several additional steps. First, we will assemble the necessary data layers for the model and scale up the pilot modeling project to cover the entire landscape. We will continue to reach out to implementers and leaders in the short term, as appropriate, including leaders within both The Nature Conservancy and the North Central Research Station. We will develop a long-term plan for building institutional support more proactively to ensure that recommendations are considered in the development of cross-boundary strategies. At the fourth forum, we plan to emphasize broad strategies, with the full complement of appropriate Nature Conservancy staff participating in the exercise.

Despite a number of early accomplishments, the success of the knowledge transfer for the project as a whole will be judged by whether an understanding of landscape ecology principles and tools ultimately influences implementation, including on-the-ground management. The desire to achieve shared management goals for the partner agencies, including the reduction of hazardous fuels, timber production, and biodiversity conservation, is at the heart of this project. Knowledge transfer of landscape ecology principles and tools during this project has been introduced as part of the toolkit that land managers and agency leaders should consider to provide a landscape context for individual ownerships. The criteria by which the initiative's long-term outcomes will eventually be evaluated lie in the answers to the following questions:

- Have the individual agencies used the products to update their own plans?
- Have the products led to more examples of strategic versus tactical collaboration among the partners?
- Are individual agencies implementing portions of the process, principles, or approach in other internal or multipartner efforts in which they are involved?
- Have individual agencies assumed ownership of the products and tools produced during this project, to the point that the origins of these aids and the initial core team have been largely forgotten?

- Has public perception changed such that there is local community support for using prescribed fire as a management tool?

Answers to these questions will likely remain unknown in the near future, but it may be possible to address them within a decade.

5.8. LESSONS LEARNED

We learned a number of key lessons during the first 2 years of the project both from our early successes and from the continuing challenges. First and foremost, we learned that building relationships was an important precursor to knowledge transfer and that relationships among agency partners were strengthened by the knowledge transfer process. Better described as a collaborative learning process, knowledge transfer in our case consisted of a regular exchange of information and ideas among landscape ecologists, resource managers, and decisionmakers. Although landscape ecologists held the keys to the modeling technology that were central to the process, land managers and decisionmakers provided crucial insights and reality checks without which even the best of technologies would have no real-world significance. Among the greatest successes was the use of spatially explicit models to build a common foundation among the partners, a step that can lead to a common vision for the landscape. Among the continuing challenges that have surfaced thus far, the most pressing are time constraints and communication among the agency partners. As a result, the process has been driven by a core team of collaborators external to the management agencies. Although key agency technical staff have been successfully engaged, the work of communicating with agency decisionmakers is only about to begin, and there is no formula yet for how to do this successfully. A short summary of the lessons learned is presented in Table 5.4 as a list of "dos and don'ts."

5.8.1. Summary of Knowledge Transfer Challenges

We learned from a number of early oversights, and were able to change course through honest feedback from team members. The overuse of landscape ecology jargon topped our list of primary things to avoid during the knowledge transfer process. It was important for the landscape scientists involved to maintain an open attitude and a willingness to learn from land managers and decisionmakers throughout the process. A successful process also required that we abandon our linear approach to landscape-level planning and adopt a more adaptive approach. Adapting our process meant respecting the need of some partners to leap ahead to strategy development at various times, allowing iterative cycles between these strategies and the desired future condition.

Initially, we failed to recognize the importance of an effective communications plan to direct communication among and within agencies so as to facilitate the development of support for the project. Likewise, internal challenges (such as those within

Table 5.4. Summary of lessons from knowledge transfer in a multipartner landscape

Do	Don't
• Provide a regular forum for knowledge transfer, such as the Fire Learning Network.	• Insist that all partners attend all knowledge transfer events.
• Ensure that the knowledge transfer forums are well organized and make good use of partner time.	• Use obscure landscape ecology jargon.
• Provide multiple opportunities for partner input, such as individual meetings, review of documents, telephone interviews, and small technical working groups.	• Take a formulaic, linear approach to landscape-level planning.
• Ensure that a core group of team leaders keeps the project moving forward.	• Allow an individual naysayer to derail the process.
• Maintain as much continuity as possible in relationships among the core team and partners.	• Overlook internal challenges to project progress, even among the core team organizations.
• Identify an internal champion within each partnering agency.	• Allow landscape ecology tools to become a "black box."
• Encourage core team members to develop support from additional key colleagues within their respective organizations.	• Forget to maintain existing relationships while building new ones.
• Use science as the foundation, and then slowly build partner relationships through knowledge transfer.	• Fail to engage key umbrella organizations early in the process (e.g., the Minnesota Incident Command System, Minnesota Forest Resources Council).

The Nature Conservancy) were initially overlooked. Although a core team was essential to this process, we initially underestimated the importance of identifying agency staff to assume leadership roles in the project and to nurture existing relationships while building new ones. Initially, we also failed to adequately engage key umbrella organizations such as the Minnesota Incident Command System and Minnesota Forest Resources Council early in the process. Their involvement has subsequently been highly important given their broad constituency and influence.

5.8.2. Summary of Knowledge Transfer Successes

We did several things well in the realm of knowledge transfer. Providing a regular knowledge transfer forum, such as the Fire Learning Network, was beneficial both for peer review and for exchange of ideas, and it helped partners to dedicate blocks of time to work exclusively on a given project. Such a forum provides structure and much-needed milestones for the project. However, for a group with such diverse, geographically scattered partners as those in the Border Lakes Partnership, it was necessary to accommodate many partners by providing separate work sessions. A combination of peer review for core leaders and smaller work sessions with key partners represented the best combination for the Border Lakes Partnership. It was crucial that the first forum was well-organized and had good content, and was the best attended of the three forums that have been held to date, because that success set the tone for future meetings.

Because we chose to use the LANDIS software, and thus needed an expert modeler to do the modeling work, it was even more critical to include all landowners in developing the modeling scenarios from the earliest stages to ensure successful knowledge transfer. The use of conceptual ecological models early in the collaborative learning process helped ensure that participants understood the natural dynamics and processes that we modeled using LANDIS. Establishing common ground at the beginning of the process assisted partners in focusing their contributions to the modeling work on key questions about forest succession, natural disturbance processes, and management activities that shape both disturbance and succession.

The importance of obtaining the assistance of an internal champion within each organization cannot be underestimated, although the degree of enthusiasm varied among these champions. Persisting in identifying a natural champion was critical, as these champions ensured that barriers continued to dissolve despite ongoing time and workload pressures. Maintaining as much continuity as possible, both among core team members and in the actual project work, and minimizing time lags were both important to the success of the project. To the extent possible, we tried to involve agency staff who expected to remain involved for the long term. In choosing a knowledge transfer approach, we learned that it is important to weigh the options of directly transferring modeling skills versus having key partners who already possess these skills guide the process.

Above all, we learned that although a science focus provides a strong foundation for such projects, knowledge transfer and collaborative learning are indispensable in building institutional support for new ideas. A successful project requires patience and a willingness to build on small successes. Knowledge transfer thus provides a bridge between the vision and the implementation that is the ultimate goal of any planning exercise.

LITERATURE CITED

Bailey RG (1995) Description of the ecoregions of the United States. USDA For Serv, Washington. Misc Publ 1391 (revised); separate 1:7 500 000 map.

Brown TN, White MA (2002) Northern Superior Uplands: a comparison of range of natural variation and current conditions. Univ Minnesota, Nat Resour Res Inst, Duluth. Interim report to Minnesota Forest Resources Council. (http://www.nrri.umn.edu/sustain/nsurnv200202.pdf)

BRW Inc (1999) GAP Stewardship (Dissolved). Minnesota Dep Nat Resour, MIS Bureau, St. Paul.

Frelich LE (2002) Forest dynamics and disturbance regimes: studies from temperate evergreen–deciduous forests. Cambridge Univ Press, Cambridge, UK.

Heinselman ML (1973) Fire in the virgin forests of the Boundary Waters Canoe Area, Minnesota. Quat Res 3:329–382.

Heinselman ML (1981) Fire intensity and frequency as factors in the distribution and structure of northern ecosystems. In: Fire regimes and ecosystem properties: Proceedings of the conference. USDA For Serv, Washington. Gen Tech Rep WO-26.

Kutas B, Doran H, Hung K, Janetos S, Strath D, Suffling R, Woodman B (2002) Management in a planning vacuum: co-operation in the Quetico–BWCA–Voyageurs boundary region. Environments 30:31–42.

MDNR (2002) GAP Land Cover—Tiled Raster. Minnesota Dep Nat Resour, For Div, Resour Assessment Unit, Grand Rapids. (http://deli.dnr.state.mn.us/metadata.html?id=L390002710606).

MDNR (2004) Draft plan: Border Lakes subsection forest resource management plan. Minnesota Dep Nat Resour, For Div, Resour Assessment Unit. (http://www.dnr.state.mn.us/forestry/subsection/borderlakes/plan.html)

Mladenoff DJ, He HS (1999) Design, behavior and application of LANDIS, an object-oriented model of forest landscape disturbance and succession. In: Mladenoff DJ, Baker WL (eds) Advances in spatial modeling of forest landscape change: approaches and applications. Cambridge Univ Press, Cambridge, UK, pp 125–162.

Mladenoff DJ, Host GE, Boeder J, Crow TR (1996) LANDIS: a spatial model of forest landscape disturbance, succession, and management. In: Goodchild MF, Steyaert LT, Parks BO, Johnston C, Maidment D, Crane M, Glendinning S (eds) GIS and environmental modeling: progress and research issues. GIS World Books, Fort Collins.

OMNR (2002) Forest management units. Data supplied under license by Ont Min Nat Resour, Peterborough. Natural Resource Values Information. http://www.lio.mnr.gov.on.ca/informationdirectory.cfm

OMNR (2003a) Land ownership. Data supplied under license by Ont Min Nat Resour, Peterborough. Natural Resource Values Information. http://www.lio.mnr.gov.on.ca/informationdirectory.cfm

OMNR (2003b) Patent land. Data supplied under license by Ont Min Nat Resour, Peterborough. Natural Resource Values Information. http://www.lio.mnr.gov.on.ca/informationdirectory.cfm

OMNR (2004a) Forest management planning manual for Ontario's Crown forests. Ont Min Nat Resour, Sault Ste. Marie. Tech Series.

OMNR (2004b) Ontario Indian reserves. Data supplied under license by Ont Min Nat Resour, Peterborough. Natural Resource Values Information. http://www.lio.mnr.gov.on.ca/informationdirectory.cfm

Pedynowski D (2003) Prospects for ecosystem management in the crown of the continent ecosystem, Canada–United States: survey and recommendations. Conserv Biol 17(5):1261–1269.

RMRS (1999) Coarse-scale spatial data for wildland fire and fuel management. USDA For Serv, Rocky Mountain Res Stn, Prescribed Fire and Fire Effects Res Work Unit, Boise. (http://www.fs.fed.us/fire/fuelman)

SMFEPT (2002) The Superior mixed forest ecoregion: a conservation plan. Superior Mixed Forest Ecoregional Planning Team, The Nature Conservancy, Madison.

Spectranalysis Inc (1999) Ontario land cover data base. Data supplied under license by Ont Min Nat Resour, Peterborough.

TNC (2004) Border Lakes Partnership: a pilot project in timber management, hazard fuel reduction, and biodiversity in a multipartner landscape. The Nature Conservancy, Arlington. Unpublished fact sheet.

USDA Forest Service (1986) Land and resource management plan—Superior National Forest, May 1986, with amendments August 2000. USDA For Serv, Milwaukee.

USDA Forest Service (2001) Final environmental impact statement: Boundary Waters Canoe Area Wilderness fuel treatment. Vol I. USDA For Serv, Milwaukee.

USDA Forest Service (2004) Land and resource management plan: Superior National Forest. USDA For Serv, Milwaukee.

Ward PC, Tithecott AG, Wotton BM (2001) Reply—a re-examination of the effects of fire suppression in the boreal forest. Can J For Res 31:1467–1480.

White MA, Host GE (2000) Mapping range of natural variation ecosystem classes for the northern Superior uplands: draft map and analytical methods. Univ Minnesota, Duluth. NRRI Tech Rep NRRI/TR-2000/39.

6

Applications of Forest Landscape Ecology and the Role of Knowledge Transfer in a Public Land Management Agency

Lisa J. Buse and Ajith H. Perera

6.1. Introduction
6.2. Applications of Forest Landscape Ecology in Ontario
    6.2.1. Sociopolitical Drivers
    6.2.2. Forest Management Policies and Guides
    6.2.3. Applications of Forest Landscape Ecological Knowledge
    6.2.4. Enabling Structures
6.3. Transfer of Forest Landscape Ecological Knowledge
    6.3.1. Users of Forest Landscape Ecological Knowledge
    6.3.2. Role of Knowledge Transfer
6.4. Applying Knowledge Transfer Principles: A Case Study
    6.4.1. Brief Description of the Study
    6.4.2. Audience
    6.4.3. Designing the Study
    6.4.4. Implementing the Study
    6.4.5. Sharing Results
    6.4.6. Disseminating Findings
    6.4.7. Developing Applications
6.5. Insights on the Transfer of Landscape Ecological Knowledge
    6.5.1. Challenges
    6.5.2. General Lessons for Landscape Ecologists

LISA J. BUSE and AJITH H. PERERA • Ontario Ministry of Natural Resources, Ontario Forest Research Institute, 1235 Queen St. E., Sault Ste. Marie, ON P6A 2E5, Canada.

6.6. Conclusions
    Literature Cited

6.1. INTRODUCTION

Over the past two decades, the application of forest landscape ecological principles has gradually become an integral part of the operations of many public land management agencies in North America. The Canadian province of Ontario is a good example of an administration where forest management policies, guides, and practices include landscape ecological principles. In this chapter, we describe why this particular region has been successful in integrating these principles into forest management, as well as how this integration occurred. In particular, we discuss the role of knowledge transfer in this process.

The province of Ontario spans more than 1 million km^2, of which more than 80% is forested. Responsibility for managing the province's natural resources rests with the Ontario Ministry of Natural Resources (OMNR), a public land management agency that is also the principal steward of nearly 0.5 million km^2 of forest land that is managed extensively for multiple values. This area is unique for two reasons: the forest cover is largely contiguous, with minimal interruption by settlements or agriculture, and nearly 90% is public land. For forest management purposes, the area is divided into 47 large planning units that range in size from 0.12 million ha to 1.56 million ha (Figure 6.1). Each unit is managed under its own management plan, which spans 20 years and is revised at 5-year intervals. Although forest management policies are developed by OMNR, operational management of forest resources occurs through long-term leases granted to private forestry companies.

As was the case for much of Canada until the 1980s, Ontario's forest management focused primarily on the production of timber and wood fiber. A combination of a global policy change to embrace sustainability, public pressure in favor of more holistic ecosystem-based management approaches, and concerns about conserving biodiversity resulted in a major shift away from the former focus on timber production. In the late 1980s, the focus also changed from managing individual stands to considering the bigger picture—the whole system and the interrelationships between its components—and this helped move Ontario away from its traditional focus on timber to a focus on larger-scale issues and approaches, which in turn led to the need for landscape-level knowledge and tools.

The single ownership, vast extent, and contiguity of Ontario's forests make management naturally conducive to larger-scale management approaches and applications. The fact that research, policy development, knowledge transfer, and operational practice are all administered by the same organization offers OMNR significant advantages in the transfer of awareness, knowledge, and skills. However,

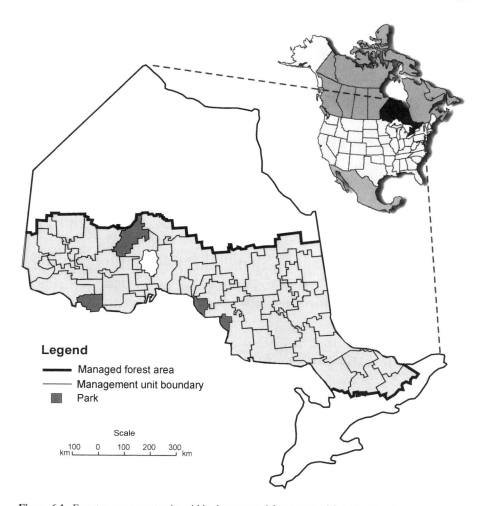

Figure 6.1. Forest management units within the managed forest area of Ontario, Canada.

additional research and practice are external to the organization, and the overall organizational structure (a public agency that develops the policies working with private companies that implement them) complicates the application of new knowledge. Thus, the process of increasing awareness, acceptance, adoption, and implementation of new concepts requires OMNR to engage a variety of audiences with very different needs and perspectives. This, in turn, requires an infrastructure that supports effective knowledge transfer.

The goal of this chapter is to examine the progress in adopting a landscape ecological perspective within a public land management agency, with an emphasis on the role of knowledge transfer in the process. Specifically, we:

- summarize where forest landscape ecological knowledge is embedded in OMNR's policies and management directions and, thus, where it is being implemented in Ontario
- examine the sociopolitical drivers and supporting infrastructure that helped to ensure that the available knowledge was adopted and applied in practice
- outline the role of knowledge transfer, and
- identify some general lessons that may help other organizations to advance the adoption and use of landscape ecological knowledge and tools in policy, planning, and practice

To illustrate our points, we provide specific examples of what has been transferred, and to whom, and explore why the concepts and tools have been adopted and are being applied. Because the choice of possible examples is large, we highlight examples with which we have firsthand experience wherever possible, including the description of an ongoing research study in which knowledge transfer was integrated from the outset.

6.2. APPLICATIONS OF FOREST LANDSCAPE ECOLOGY IN ONTARIO

Ontario's forest policy framework encompasses societal, economic, and ecological values that are addressed within global, national, and local contexts. The framework is organized into levels—strategic directions, legislative and regulatory requirements, provincial policies, and strategies—that are expressed in a series of forest management guides that direct planning and practice, as well as operational and administrative directions. As Euler and Epp (2000) pointed out, this framework is designed to allow periodic adaptation of policies and guidelines through regular reviews and revisions that permit the incorporation of new knowledge. All levels of this framework are informed by forest landscape ecological principles, thus providing a continuous link from legislation to policy and from policy to practice.

In this section, we explore some of the drivers that motivated recipients to embrace the new concepts, provide examples of where forest landscape ecological knowledge is embedded in both policy and practice in Ontario, and outline the enabling factors that supported the adoption of landscape-level approaches. First, we identify several factors that enabled Ontario to successfully adopt a landscape ecological perspective. These include an increased focus on sustainability and biodiversity worldwide, and a number of concurrent sociopolitical drivers at the local level, all of which were aided by the expanding global and local knowledge base in landscape ecology.

Colour Plates

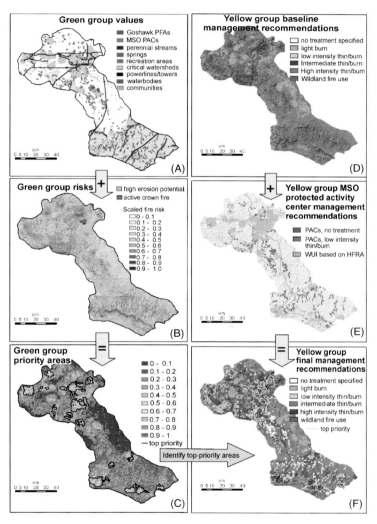

Figure 4.3. Two parallel processes for using (A) values and (B) risks to (C) prioritize areas and specify treatments (D and E) are combined to develop a sample management action scenario (F). The Green group was one of four breakout groups at the first Western Mogollon Plateau Adaptive Landscape Assessment workshop held in February 2004 who developed prioritization maps. The Yellow group, made up of different stakeholders, was one of four breakout groups at the second workshop held in May 2004 who developed management recommendations. HFRA, Healthy Forests Restoration Act of 2003; MSO, Mexican spotted owl; PACs, protected-activity centers; PFA, postfledging areas; WUI, wildland–urban interface.

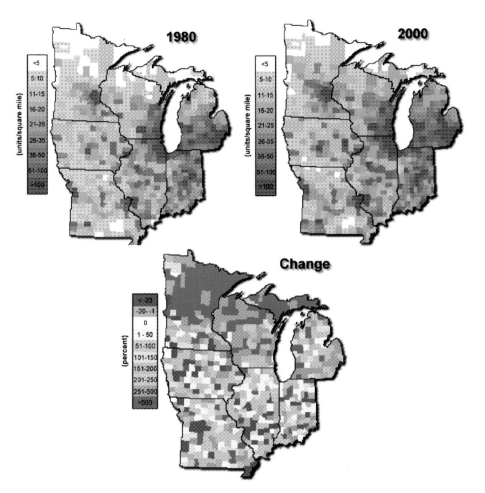

Figure 7.2. Changes in housing density between 1980 and 2000 in the north-central region of the United States. Major increases in housing density in the upper part of the region have occurred without major increases in human population because many new sites represent second homes for people already living in major metropolitan areas. Source: The Changing Midwest Assessment (http://ncrs.fs.fed.us/4153/deltawest/).

6.2.1. Sociopolitical Drivers

Global drivers. In the early 1980s, a global shift occurred in public awareness of the concept of biodiversity, leading to perceptions that forests comprise more than just trees and provide more values than just timber, and that biodiversity cannot be conserved solely at the scale of forest stands (Brundtland 1987). The shift within OMNR from timber management to broader-scale sustainable forestry occurred in the early 1990s, driven by these trends in conservation of biodiversity and concerns about overall forest health and sustainability. Epp (2000) provides a detailed chronology of how these events related to Ontario. More recently, the global trend in favor of third-party certification of forestry operations has required the forest industry to conserve biological diversity and associated values, water resources, soils, and unique and fragile ecosystems and landscapes. By so doing, the industry would maintain the ecological functions and integrity of the forest. As of September 2005, almost 27 million ha of managed forest area in Ontario (more than 50% of the total) were certified under one or more certification systems (Certification Canada 2006). As the requirements associated with maintaining certification evolve, they will continue to pressure forest companies to consider landscape dynamics and functions in their management practices.

Local sociopolitical drivers. In the late 1980s, an Ontario-wide environmental assessment of timber management practices (a Class Environmental Assessment) was undertaken (OEAB 1994), and this exercise provided the impetus for a series of changes in forest policy, and the recognition of the need for a series of forest management guides to provide direction during forest management planning. Many of the resulting guides focused on managing the supply of wildlife habitat for animals that require large or diverse areas (e.g., pine marten, *Martes americana*; woodland caribou, *Rangifer tarandus caribou*; moose, *Alces alces*; red-shouldered hawk, *Buteo lineatus*), with the associated knowledge encapsulated in specific directions (see Table 6.1b for examples of such guides). Creation of these guides required landscape ecological knowledge to provide context, mostly from sources outside the provincial government, and an additional push from researchers. This process coincided with the evolution of an early-1990s sociopolitical policy program, the Sustainable Forestry Initiative, that led to the creation of the Policy Framework for Sustainable Forests (OMNR 1994); this framework provided the overall context for forest management in Ontario and, most importantly, led to the development of a new forestry act (the Crown Forest Sustainability Act; Statutes of Ontario 1995) that entrenched forest sustainability at the legislative level. Euler and Epp (2000) provide considerable insight into the development of these directions.

6.2.2. Forest Management Policies and Guides

The Crown Forest Sustainability Act (Statutes of Ontario 1995) provides an overarching legislative direction for Ontario's management of public forest land and addresses the value of emulating natural landscape disturbance to conserve

Table 6.1a. Examples of where forest landscape ecological principles and knowledge have been incorporated into Ontario's major forestry legislation, policies, and planning directions

Title (reference)	Forest landscape ecological principles, knowledge, and directions
Legislation	
Crown Forest Sustainability Act (Statutes of Ontario 1995)	• Emulate natural disturbances and landscape patterns to sustain forests and conserve biodiversity • Ensure sustainable forest management
Policies and strategies	
Policy Framework for Sustainable Forests (OMNR 1994)	• Maintain ecological processes • Conserve biological diversity • Emulate natural disturbances and maintain landscape patterns
Ontario's Land Use Strategy (OMNR 1999)	• Expanded Ontario's provincial parks and network of protected areas to include a representative spectrum of Ontario's ecosystems and natural features (identified by ecoregion) • Protect and manage entire watersheds

Old Growth Policy (OMNR 2003)	• Provides a landscape management perspective for the conservation of old growth
	• Directs resource managers to:
	  • use spatial simulation modeling to assess current and future abundance and distribution of old-growth forests based on succession and natural disturbance patterns
	  • acknowledge the spatial and temporal dynamics of old-growth forests
	  • set targets for old-growth forests based on the probabilities of aging, occurrence, and distribution within ecoregions

Ontario Biodiversity Strategy (OMNR 2005)	• A landscape approach to biodiversity conservation
Planning directions	
Forest Management Planning Manual (OMNR 1996, 2004)	Directs forest managers to: • Assess landscape pattern indices as indicators of biodiversity in relation to provincial, regional, and subregional levels • Document current and future availability of wildlife habitat in provincial, regional, and subregional contexts • Document net primary productivity and water yield as indicators of landscape processes • Project forest succession and disturbance rates for 150 years

biodiversity and enhance the sustainability of forests. Several subsequent policies and strategies, such as Ontario's land-use strategy (OMNR 1999), contain further landscape ecological concepts, such as the need to maintain ecological processes and consider spatial and temporal variation at broad scales. Ensuing directions for forest management planning and guides to support forest management, all of which help to operationalize policies and strategies, contain specific landscape-level applications, such as using natural disturbance templates customized for each ecoregion to design spatiotemporal harvesting patterns, providing wildlife habitat at levels ranging from forests to ecoregions, and monitoring spatial heterogeneity and ecological processes in the managed forest. Table 6.1a summarizes where forest landscape ecological knowledge is embedded within Ontario's policy framework. For a complete description

A Public Land Management Agency

Table 6.1b. Examples of where forest landscape ecological principles and knowledge have been incorporated into Ontario's forest management guides

Title (reference)	Forest landscape ecological principles, knowledge, and directions
Forest management guides	
Timber Management Guidelines for the Provision of Moose Habitat (OMNR 1988)	• The first forest management guide to consider spatial patterns of cut blocks (size, distribution, edges) over larger areas with respect to the effects on wildlife habitat
Forest Management Guidelines for the Provision of Marten Habitat (Watt et al. 1996)	• Identifies the need to consider landscape-level effects of management practices on overall habitat availability as part of forest management planning • Provides guidelines for maintaining landscape composition, patterns, and structure to benefit pine marten populations • Recommends use of habitat supply models to ensure that sufficient preferred habitat remains across management units
Forest Management Guidelines for the Provision of Pileated Woodpecker Habitat (Naylor et al. 1996)	• Identifies the need to consider landscape-level effects of management practices on overall habitat availability as part of forest management planning • Recommends use of habitat supply models to ensure that sufficient preferred habitat remains across management units
Forest Management Guidelines for the Provision of White-Tailed Deer Habitat (Voigt et al. 1997)	• Identifies the need to consider landscape-level effects of management practices on overall habitat availability as part of forest management planning • Recommends use of habitat supply models to ensure that sufficient preferred habitat remains across management units in each season
Forest Management Guidelines for the Conservation of Woodland Caribou—A Landscape Approach (Racey et al. 1999)	• Recommends that caribou be managed on very large spatial and temporal scales (i.e., spanning more than a single management unit over more than 80 years) • Directs that management decisions be supported by analyses of spatial habitat supply to ensure that sufficient contiguous forest is provided
Forest Management Guide for Natural Disturbance Pattern Emulation (OMNR 2002)	• Provides standards and guidelines for emulating natural (fire) disturbance patterns when harvesting forests

of Ontario's forest policy and legislative framework, see the Ontario's Forests Web site (http://ontariosforests.mnr.gov.on.ca/ontariosforests.cfm).

Over the next decade, new policies and guides were founded on the emulation of natural disturbance regimes, with the focus changing from fragmentation and habitat issues to biodiversity and conservation issues, driven in part by the Crown Forest Sustainability Act, which specified that forest management be based on emulating natural disturbance and landscape patterns. One example of a current large-scale provincial policy is the *Forest Management Guide for Natural Disturbance Pattern Emulation* (OMNR 2002), which directs forest managers to move toward natural landscape patterns; matching management approaches to what *could* happen rather than *what once happened* requires an assessment of larger-scale, longer-term landscape dynamics, and an understanding of the potential variation in disturbance patterns. The evolution of this policy is documented in detail by McNicol and Baker (2004). Another such policy is the *Old Growth Policy* (OMNR 2003), which requires larger-scale, longer-term thinking about how much existing old-growth forest should be conserved and where to plan for the future development of old-growth forest, which is not a static entity. Both the conservation of old-growth forest and the emulation of natural disturbance patterns were built into the directions provided in the provincial Forest Management Planning Manual (OMNR 1996, 2004). The land-use planning process has also incorporated landscape-level approaches. For example, a provincewide exercise conducted in the late 1990s incorporated an ecoregion-based land-use planning hierarchy for managing forest landscapes as well as watersheds and protected areas. Francis (2000) provides more details about strategic land-use planning in Ontario. More recently, a review of the existing forest management guides (AES et al. 2000) led to a consolidation of the existing documents into a concise set of five guides, one of which addresses topics explicitly at landscape scale and integrates forest landscape ecological knowledge and applications in a single document.

6.2.3. Applications of Forest Landscape Ecological Knowledge

As landscape ecological concepts were being incorporated into forest management policies, practitioners were faced with the challenge of implementing the policies in their planning and practices. To do this, they needed to understand the concepts (which required knowledge transfer) and a means to implement the policies (e.g., by providing tools). Thus, there has been a push both to increase awareness and knowledge of landscape ecological concepts and to develop tools that can help practitioners apply the new concepts embedded in the policies and guides. This required transfer of the relevant skills and knowledge that would enable practitioners to use the tools and to interpret and apply the output of the tools to achieve the desired management outcomes. For example, Ontario's management unit–level forest management plans require the application of landscape ecological principles, such as analysis of landscape patterns (e.g., connectivity, patch size), analysis of habitat supply, and monitoring of changes in primary productivity at local and regional levels.

User applications have been developed and revised to support these needs, in part because the policies created a need and in part because practitioners demanded an efficient and effective way of getting the information they needed to meet the new requirements (Table 6.2). These include tools for assessing landscape patterns (LEAP II, Perera et al. 1997; Patch Analyst, Elkie et al. 1999b; NDPEG Tool, Elkie et al. 2002), landscape processes (RHESSys, Band 1993; ON-FIRE, Li et al. 1996; BFOLDS, Perera et al. 2004; NPPAS, Schnekenburger and Perera 2003), habitat supply (Ontario Marten Analyst, Elkie et al. 1999a; OWHAM, Naylor et al. 2000), and landscape-level harvest planning (SFMM, Kloss 2002; Patchworks, SPS 2006). Most of these were initially used as research models or tools, but have since been transformed or are in the process of being transformed into desktop tools that can be used

Table 6.2. Examples of forest landscape ecological tools and applications developed to support forest management planning in Ontario

Tool or model	Name	Purpose	Description	User group
RHESSys (Band 1993)	Regional hydro-ecological simulation system	Quantify carbon, water, and nutrient fluxes	Simulates the spatial distribution and spatial and temporal interactions of carbon, water, and nutrient fluxes at the watershed scale.	Regional analysts
LEAP II (Perera et al. 1997)	Landscape ecological analysis package	Quantify landscape patterns	Supports the calculation of landscape metrics such as patch area, density, edge, nearest neighbor, diversity, and interspersion for various forest classification schemes.	Regional analysts and forest management planning (FMP) teams
Patch analyst (Elkie et al. 1999b)	Ontario Patch Analyst	Quantify landscape patterns	Supports the calculation of landscape metrics such as patch area, density, edge, nearest neighbor, diversity, and interspersion for various forest classification schemes.	FMP teams
OMA (Elkie et al. 1999a)	Ontario Marten Analyst	Assess habitat availability	Allows spatial assessment and classification of habitat based on forest resource inventory data through time; used primarily for identifying core pine marten areas and caribou habitat in northwestern Ontario.	Regional analysts and FMP teams
OWHAM (Naylor et al. 2000)	Ontario wildlife habitat assessment model	Assess habitat availability	Allows spatial assessment and classification of habitat based on forest resource inventory data through time. Used to identify habitat for moose, caribou, deer, red-shouldered hawk, and pileated woodpecker in northeastern and central Ontario.	Regional analysts and FMP teams
NDPEG tool (Elkie et al. 2002)	Natural disturbance pattern emulation guide tool	Analyze landscapes to support emulation of natural patterns	Summarizes historical fire data and analyzes current and future landscape disturbance patterns based on rules specified in the *Forest Management Guide for Natural Disturbance Pattern Emulation* (OMNR 2002).	Regional analysts and FMP teams
BFOLDS (Perera et al. 2003)	Boreal forest landscape dynamics simulation model	Simulate landscape dynamics over large areas and long time frames	A spatially explicit model that simulates natural disturbance and succession. Can be used to explore and understand the spatial, temporal, and random variations in fire disturbance regimes in boreal forests.	Regional analysts

by practitioners. Knowledge transfer initially comprised explanations of the models and guidance in interpretation of the results, but evolved into ongoing training on how to use the tools as their development progressed. The development and transfer of these tools were made possible by local generation of knowledge and the existence of an adequate technological infrastructure (e.g., sufficient computing power).

6.2.4. Enabling Structures

Several factors contributed to successful adoption of forest landscape ecological knowledge in Ontario. As outlined above, policies informed by landscape ecological concepts were developed in response to a combination of global and local drivers based on increased interest in sustainable forestry and biodiversity, which in turn created a niche for tools and databases to support the implementation of these policies. In essence, the development of knowledge pushed the development of a policy framework and the ensuing demand from forest resource managers pulled the development of more knowledge in the form of tools and knowledge to help implement the policies. This knowledge transfer process was enabled by Ontario's capacity for the generation of local knowledge and OMNR's organizational and supporting infrastructure.

Ontario benefits from a continuum of developers of basic and applied landscape ecological knowledge. These developers include researchers at 13 universities, the federal forest service, and a research branch within OMNR, all of whom generate landscape-level information and tools. This capacity has generated a considerable forest landscape ecological knowledge base in the primary and secondary literature, beginning in the late 1980s. This is evident in a recent compilation of research knowledge on the forest landscape ecology of Ontario (Perera et al. 2000). Moreover, OMNR has established a geographically dispersed network of science and information units in which science specialists support the transfer of global landscape ecological concepts to produce local applications and tools for forest resource managers.

In addition to knowledge generation and transfer capacity, an enabling infrastructure also supported the implementation of landscape ecological knowledge and tools, along with the necessary organizational resources. Extant Ontario-wide spatial databases include several sources of periodically updated data on forest cover (e.g., Landsat TM, airphoto-based forest resource inventory) and forest disturbances (e.g., harvesting, fire, insect epidemics); geographic information system (GIS) climatic and geological databases (e.g., soils, geology, climate, terrain, watersheds); databases on species habitats, wetlands, and other environmentally sensitive areas; and ancillary spatial information (Table 6.3). This array of readily accessible GIS databases made the practical application of landscape ecological tools feasible both at an Ontario-wide level for policy development and at the level of management units to support the management of forest resources. Simultaneously, fueled by the global growth in information technology, Ontario-wide networks and standards were established so that GIS databases, software, and

A Public Land Management Agency

Table 6.3. Ontario-wide GIS databases available to support the development of forest policy, planning, and management

Category	Spatial database	Description
Forest cover	Airphoto-based forest resource inventory	Species and age composition and stand characteristics; updated every 10 years (1:20 000)
	Landsat TM forest cover classification	Broad forest-type classes; updated every 5 years (30-m resolution)
Physiography	Ontario Land Inventory Northern Ontario Engineering and Geology Terrain Survey	Broad soil groups, moisture capacity, nutrient regime, and other pedological characteristics (1:250 000 to 1:50 000)
	Ontario surficial geology atlas	Combination of terrain composition and spatial distribution of glacial geological materials (1:250 000 to 1:500 000)
	Ontario digital terrain model	Slope, aspect, and ruggedness (100-m horizontal and 5-m vertical resolution)
Disturbance history	Forest fire history	Burned area, dates of burns, and suppression activities (1:20 000), updated annually
	Forest harvest history	Harvest patches and residual areas (1:20 000), updated annually
	Ontario forest disturbance survey	Biotic and abiotic causal factors such as insect pests and windthrow (1:50 000), updated annually

hardware systems were compatible among knowledge developers and users. This ensured that any tools developed by researchers would be accessible and applicable in every forest management unit in Ontario. As a result, the practical obstacles encountered elsewhere in applying landscape ecology—a lack of data and the unavailability of appropriate computing technology—were not impeding factors in Ontario. Ontario's implementation of GIS-based planning and management approaches in the 1990s required that trained people be in place to ensure the successful transfer and use of new forestry applications of this technology. Beginning as early as 1992, all OMNR offices and the forest industry acquired personnel with GIS expertise, along with the supporting information technology, and this helped to ensure that landscape-level tools and databases could be used in all forest management organizations. The efforts to ensure compatibility of GIS data, systems, and skills across Ontario considerably facilitated the adoption of this technology and provided a unique avenue for the transfer and application of landscape ecological tools. However, the transfer of landscape ecological concepts and knowledge to policymakers and practitioners still had to occur.

6.3. TRANSFER OF FOREST LANDSCAPE ECOLOGICAL KNOWLEDGE

Transferring research knowledge and tools into practical use requires that knowledge developers know who their users are and actively engage them. In this section, we describe the primary users of forest landscape ecological knowledge in Ontario and provide a brief overview of how the province's knowledge development and transfer structure supports the transfer of knowledge to these audiences.

6.3.1. Users of Forest Landscape Ecological Knowledge

The hierarchy of Ontario's policy framework and the shared responsibility for forest management between the forest industry and OMNR make the community of users of forest landscape ecological knowledge both diverse and complex. We identify three broad user groups, each of which requires and uses different aspects of landscape ecological knowledge: decisionmakers, policymakers, and forest resource managers. These groups exist both in the public sector and in private forestry companies.

Decisionmakers include those who shape legislative directions in the public sector and broad-scale forest management strategies for forestry companies in the private sector. This group is receptive to landscape ecological concepts and their applications that are relevant to global and national forest management issues such as climate change, invasive species, conservation of endangered species, and forestry certification. They are also responsive to Ontario's socioeconomic and political milieu, and are responsible for incorporating landscape ecological concepts in several broad forest management strategies. However, decisionmakers such as politicians, leaders of public sector agencies, and forest industry executives receive this knowledge through their advisory staff, making the latter individuals the most direct users of broad-scale landscape ecological knowledge and therefore the direct recipients of knowledge transfer. These advisors typically have academic backgrounds and some practical experience in forestry or wildlife biology, and are interested in answers to the *what* and *why* of landscape ecological knowledge with respect to forest management.

Policymakers are also interested in landscape ecological concepts, but more in relation to forest management applications—that is, they serve as a bridge between broad-scale directions defined by decisionmakers and the actual practice of forest management by practitioners. This group belongs exclusively to the OMNR, the primary public sector forest management organization in Ontario. Ontario's forestry policymakers have a diverse array of academic backgrounds including biology, forestry, and land-use planning. They understand biological and ecological principles and commonly have practical experience in forest or wildlife management. Policymakers have the advantage of being in direct contact with both forest managers and knowledge developers. One challenge they face in developing new policies and guides that incorporate forest landscape ecological concepts is how to

A Public Land Management Agency

balance broad socioeconomic realities with the larger-scale and longer-term scope of landscape ecology. Knowledge developers have interacted with and continue to interact with policymakers to assist in the policy development process. Ontario's policymakers have been responsible for a series of forest management policies and guides that contain landscape ecological knowledge (outlined in Tables 6.1a,b).

Forest resource managers include professionals who plan and implement forest management operations under Ontario's forest policy framework; all are trained in forestry, wildlife biology, or related disciplines, with at least some level of university or college education. Forest resource managers in the public sector (OMNR) assist in developing and approving forest management plans, whereas those working for private forestry companies are responsible for developing and implementing those plans. Therefore, their use of landscape ecology is mainly focused on landscape ecological applications such as tools and databases. Although many are interested in the underlying landscape ecological concepts, and some pursue information beyond what is required to use the tools, their main focus remains the results of using the transferred knowledge and tools and their applicability in the context of the socioeconomic and short-term realities that constrain forest management. Knowledge developers and transfer specialists interact with forest resource managers by various means: presentations, workshops, training sessions, and one-on-one discussions. The proportion of forest resource managers who have not had an opportunity to learn landscape ecology and related spatial and GIS techniques has traditionally been high. Consequently, an intermediate user group, consisting of GIS technologists, has evolved as a necessary component in the process of applying the tools of landscape ecology. In contrast to the focus of decisionmakers and policymakers on the *why* and *what* of landscape ecology, this user group is focused on the *how* (i.e., on the practical and applied uses of landscape ecology in forest management).

Two other broad groups of indirect users of landscape ecological knowledge hold considerable influence in forest management in Ontario: case-specific stakeholders (e.g., other users of forested land, environmental nongovernmental organizations) and the general public. Although these groups have been instrumental in shaping some of the province's forest policies and management practices, we have not included them in our discussion in this chapter because they have no direct responsibility for management and thus are not primary (direct) recipients of the knowledge developer's transfer efforts.

6.3.2. Role of Knowledge Transfer

Ontario's capacity for knowledge transfer and its approach to transfer are unique. Ontario's universities do not have forest extension programs as are common in American universities. As well, unlike the federal forest service in the United States, Canada's federal forest service is not responsible for formal extension to forest managers or provincial policy developers—its transfer focus is national policy development. Therefore, the primary responsibility for knowledge transfer in Ontario rests with OMNR, specifically knowledge developers and transfer specialists. OMNR has three groups of terrestrial knowledge developers, all with research capacity in landscape ecology, and three transfer specialist groups that are geographically dispersed to match local forest landscape characteristics.

The knowledge developers are primarily researchers who generate landscape ecological knowledge in much the same way as academic researchers at a university, but are focused on applied problem-solving related to Ontario's policies and practices. Up to half of their time is devoted to actively participating in knowledge transfer. For example, in addition to generating new knowledge, such as models and research findings, and publishing the information in journals and books (as is customary for researchers), these individuals also are expected to produce material that is directly usable by policymakers and resource managers, such as user manuals and user-friendly software tools. The transfer specialists are science professionals, generally with some practical management experience, who adapt the knowledge created by the developers to meet the resource manager's needs and who provide training on associated concepts and tool use to resource managers.

For example, as part of regular training in forest management planning, these transfer specialists have developed an intensive training module specifically designed to introduce landscape-related concepts such as old-growth forests, biodiversity, and wildlife habitat assessment, and to train practitioners to use relevant tools (e.g., those outlined in Table 6.2) while developing management plans. This module became a means of providing ongoing transfer of existing and new knowledge and tools to forest management planners in OMNR and the forest industry on a 5-year cycle, with input from knowledge developers, policymakers, and practitioners. Ideally, developers and transfer specialists work together to transfer knowledge and applications by developing products together or through joint transfer efforts. As an example, the provincial landscape guide currently being developed involves a team approach in which a development team that includes forestry companies and stakeholders advises OMNR policymakers on the scope, content, and implementation of the guide, and two scientific teams that include knowledge developers, transfer specialists, and policymakers support the development team. These teams work together to make predictions and explore extremes in the possible outcomes of policy alternatives and in doing so, increase their understanding of the associated concepts and how the model works. This in-person, hands-on transfer requires dedicated, knowledgeable, and trained individuals, and a planned, yet flexible approach in which the landscape ecological concepts and rationale are introduced before and in conjunction with training in the use of models and tools. We briefly outline how this worked for two tools, one that has been successfully transferred to users (the OWHAM–OMA combination) and one that is in the process of being transferred (BFOLDS).

Wildlife Habitat Assessment Models

The Ontario Wildlife Habitat Assessment Model (OWHAM), which is used in the central and northeastern regions of Ontario, and the Ontario Marten Analyst (OMA), which is used in the northwestern region, both allow spatial assessment and classification of habitat through time based on local forest inventory data and knowledge of ecological succession. They evolved from a need identified by planning teams attempting to apply forest management guides for the provision of habitat for wildlife, including white-tailed deer (*Odocoileus virginianus*; Voigt et al. 1997), the pileated woodpecker (*Dryocopus pileatus*; Naylor et al. 1996), and the pine marten (Watt et al. 1996). For example, OMA was developed following requests for a method to analyze habitat availability and identify core pine marten areas within forest management units in a clear, consistent, and transparent manner. These models were transferred through a combination of presentations and training workshops and are now used by forest management planning teams in both government and industry.

The Boreal Forest Landscape Dynamics Simulator

The BFOLDS model simulates the boreal forest's fire disturbance regime and succession at the level of ecoregions (several millions of hectares) over time spans of several centuries. It is currently being used to simulate the probabilities of forest fire and forest-cover transition scenarios to provide benchmark information for the development of OMNR's landscape guide for forest management. In addition, it serves as a means to transfer the principles of longer-term variation in potential disturbance patterns, and to explore and understand the nature of the boreal forest's fire disturbance regimes and how these vary through time and over large areas. This knowledge provides insights into how resource managers can influence future forest landscape conditions in a spatially explicit manner. Model transfer has occurred through presentations that provide step-by-step explanations of the process and of the concepts behind emulating natural disturbance, including the inherent variability, and that discuss the results of simulation runs. Over the past 2 years, knowledge developers have used three hands-on training workshops, numerous presentations to users ranging from policymakers to forest managers, and a number of demonstration simulations using the model to facilitate transfer. As well, the professionals responsible for the development of landscape policy have been involved in calibrating the model and performing sensitivity-analysis simulations using local data and expertise. This interactive use of the model serves as a precursor to and as part of the training process for its future use in land-use planning at regional and management unit levels.

As the above examples demonstrate, the role of knowledge transfer in OMNR is shared among knowledge developers, transfer specialists, and policymakers, who work together to perform knowledge transfer activities that include presentations, workshops, and training in tool use in an applied problem-solving framework, leading to revised policy or improved approaches to forest management planning

A unique combination of global and local drivers, a large and expanding knowledge base, an appropriate support infrastructure, and in-house transfer capacity supported successful transfer of landscape ecological knowledge in Ontario. For the most part, the rationale (*why*) was a given that was entrenched in legislation and policies, so it was mostly the *what* (e.g., the scope and contents) and the *how to* that were transferred. Transfer included both increasing the awareness, knowledge, and skills of individuals, and enhancing policies and directions to support the development of revised management guides.

Given the drivers, the knowledge base, and the supporting infrastructure, how exactly does knowledge transfer occur? In the next section, we explore the mechanisms of knowledge transfer through an ongoing case study.

6.4. APPLYING KNOWLEDGE TRANSFER PRINCIPLES: A CASE STUDY

We are presently conducting a multiscale research study designed to increase our understanding of natural fire regimes in boreal Ontario and provide better guidance on how to emulate this form of natural disturbance through forest management. Knowledge transfer is being integrated into the study both to enhance the applicability of the research and to ensure that the intended audiences are aware of the knowledge as it becomes available. In this section, we briefly illustrate how this knowledge transfer is being accomplished within the context of the larger research project. Given that many examples of successful transfer of landscape ecological knowledge exist in Ontario, we based our choice of this example entirely on our familiarity with the project.

Following the researcher's initial concept or idea, research projects commonly progress through a series of stages: experimental design, implementation, analysis of results, and reporting the results. In some cases, and especially so for landscape ecological projects, developing applications of the knowledge and providing training in those applications follows the reporting stage. Engaging specific audiences at various stages throughout the study increases awareness not only of the project but also of the intended outcomes. We are using an integrated approach based on concurrent research and transfer of knowledge to the intended audiences (the potential users of study results) based on an ongoing discovery of their needs at various stages of the study.

6.4.1. Brief Description of the Study

This study addresses the characteristics of fire regimes at multiple scales: characteristics of the fire regime at an ecoregional scale, of fire events at a subregional scale, and of subfire events at a stand scale. In other words, we ask the following question: What patterns do fires create in a forested landscape at different spatiotemporal scales? Our study emphasizes an understanding of how and why these natural patterns vary in both time and space so that resource managers can better emulate these patterns through their management decisions. The knowledge gained by the study will be used to revise specfic forest policies and practices related to the broader policy of emulating patterns of natural forest disturbance (OMNR 2002).

6.4.2. Audience

The intended audiences for the natural fire regime study are decisionmakers, policymakers, and practitioners in boreal Ontario. For the purposes of our case study, we define these audiences as follows:

- *Decisionmakers* are those with the authority to decide research priorities and control funding. They require an understanding of the rationale for the study and how it fits conceptually with other organizational policies, directions, and priorities. Decisionmakers will also be interested in the general findings of the study and how these could be used in policy and practice.
- *Policymakers* are those who incorporate research results into resource management policies and guides. They require an understanding of the linkages and relevance of the study to specific policies, how the knowledge is being developed, where the new knowledge will be integrated into extant or new policies, the knowledge gaps that the study will and will not address, and their implications, as well as the eventual applicability of the results.
- *Practitioners* are those who will implement the policies and guides that result from the study in forest management planning and operations. They require an understanding of how the research results can assist them in solving management problems and are most interested in the tools developed to help them implement the results.

An overview of how these audiences are being engaged at various stages of our study is provided in this section to illustrate both the mechanisms being used to accomplish the transfer and the benefits of ongoing engagement with the intended audiences. At the time of writing, we have completed the study design and have begun the implementation stage. Therefore, we will describe knowledge transfer efforts with respect to what we did during the design stage, what we are doing during the implementation stage, and what we will do during subsequent stages. Table 6.4 summarizes our overall approach.

6.4.3. Designing the study

During the study design phase, the intended outcomes of knowledge transfer were awareness and engagement. To accomplish these outcomes, even before all the study details had been developed we presented an overview of the background and rationale for the study, the proposed approach, and an indication of how the findings will benefit the organization to decisionmakers and used their feedback to refine our study proposals.

Policymakers are especially interested in influencing how the research is conducted. To satisfy this need, we included several policymakers as formal study advisors; they reviewed our study proposals to critique the scope, goals, methods, and time frames of the research. Their involvement in the project was through more

Table 6.4. Overview of the knowledge transfer stages, audiences, objectives, and possible methods used at various stages of an ongoing research project on natural fire regimes in boreal Ontario

Research project stage	Transfer stage	Audience	Transfer objective and content	Transfer method
Design	Awareness Early engagement Incorporating user ideas	Decisionmakers Policymakers, Practitioners	Concepts, Rationale Concepts, Rationale, Methods	Overview presentation on the approach Technical presentation on the research design In-person discussions Joint field visits Analysis of knowledge gaps Review of research proposal
Implementation	Incorporating user ideas Maintaining engagement	Practitioners	Methods	Technical presentation on methods In-person discussions Joint field visits Technical workshops
Analysis of results	Maintaining engagement Sharing preliminary results	Policymakers, Practitioners	Concepts, Methods, Outcomes	Technical presentation on early results In-person discussions
Disseminate findings	Maintaining engagement Sharing final results	Policymakers, Practitioners	Concepts, Outcomes, Applications	Technical presentation on findings and applications In-person discussions
Develop applications	Incorporating user ideas Moving toward implementation	Policymakers, Practitioners Decisionmakers	Application tools Outcomes, Applications	Training workshops Overview presentation on general findings and potential uses

A Public Land Management Agency 147

detailed technical presentations and discussions than those aimed at the decision-makers, and this engagement allowed us to incorporate their ideas and perspectives into the study design. As a result, policymakers understand what specific uncertainties in current policies are being addressed by this research, and how.

The practitioners we contacted included foresters, biologists, resource technicians, and planners. Initially, we also informed them of the study rationale and approaches through an overview presentation and discussions, and subsequently involved them in several field visits to potential research sites. During these visits, field foresters and biologists provided feedback on the proposed study design and offered relevant local data and information that could be used to enhance the proposed study. These small group discussions also identified additional questions that should be investigated during the study.

In summary, we engaged more than 10 different audiences, ranging from decisionmakers to policymakers and practitioners, to provide an overview of the study rationale, approaches, methods, and time frames by means of presentations, meetings, and field visits. This approach ensured that many individuals belonging to various groups of knowledge users became aware of the study, accepted the proposed approaches, and understood the value of and the need for this research. These activities thus represented early knowledge transfer, and provided a user-review of the research in parallel with traditional peer review of the methods by fellow researchers.

6.4.4. Implementing the Study

During the study implementation phase, our transfer focus is on the practitioners who will be involved in discussions about the research methods and applications of the results. For example, one aspect of our study uses high-resolution aerial photography to map patterns of fire residuals (patches and trees remaining in burned areas). We are illustrating the data collection process and the objective methods of error analysis to ensure that errors and limitations of the data are clear and acceptable to those who will be applying the results. In addition, their involvement in this stage of the study is stimulating interest in the early results, and is providing opportunities for feedback. As the study progresses, we will ensure continuity in engagement with this audience through interim presentations, field visits, and sharing of the interim study results. This will help us to familiarize our audience with new technologies being used in the research project and to discuss potential challenges in applying the expected results in the field.

6.4.5. Sharing Results

We plan to transfer interim study results to policymakers and practitioners who were involved during the study design and implementation phases. Policymakers will begin thinking about how these results may fit with existing policies or what policy revisions may be justified based on our findings, while practitioners can start incorporating results into their planning and operational practice. Preliminary results are best shared through presentations and discussions that review the study rationale and methods once more so that intended applications (and any associated limitations) are clear. Questions and ideas generated during these sessions can stimulate further data analyses and help us to refine our interpretations of the results. For researchers, this step may reveal which aspects of the results will be most difficult to communicate and transfer during implementation.

6.4.6. Disseminating Findings

Once final results are available, we will disseminate them to a broader audience. Even though awareness and understanding of the concepts and approaches remains important, the transfer goal will shift toward ensuring that the new knowledge becomes embedded in new or revised policies and practices. This step will be accomplished in concert with policymakers and practitioners through technical presentations and discussions, in addition to the standard publication and distribution of reports and journal papers. Transfer initiatives and products will again include a review of the study rationale and embedded concepts, but with the focus changing from the approach to the outcomes and potential applications of results.

6.4.7. Developing Applications

The complex information that results from such a study can be built into existing applications or used to develop new applications that practitioners can use to support their planning or operational practice. This will involve working with regional planners to develop tools and associated training workshops. Training workshops for users increase their comfort with the tools, provide an opportunity to address user concerns about the tools and their application, and increase awareness of the embedded landscape ecological concepts. Once again, incorporating ideas generated by our audience will increase the likelihood that the tools and their applications will be accepted.

Once the final outcomes and applications are developed, we will reengage the decisionmakers to present the general findings and potential uses of the knowledge. This step serves to reinforce the relevance of the study, creates awareness of how and where the results are being or may be used, shows linkages to other organizational needs, and relates the results to future research needs. This keeps decisionmakers informed of relevant advances and closes the knowledge transfer loop.

6.5. INSIGHTS ON THE TRANSFER OF LANDSCAPE ECOLOGICAL KNOWLEDGE

In this section, we summarize our experiences in Ontario over the past two decades to offer insights for developers of landscape ecological knowledge. Although we focus primarily on successes in knowledge transfer throughout the chapter, we also encountered many challenges in the transfer of landscape ecological knowledge in Ontario.

6.5.1. Challenges

Our experiences suggest a complex assortment of impediments to successful knowledge transfer: the problems may be transient, temporary, or long term; case-specific or systemic; limited or pervasive; and caused by individual personalities or organizational culture. Some of these challenges may be minimized by effective knowledge transfer.

- *Unfamiliarity with landscape ecology.* The traditional stand-level educational background of most resource managers makes them focus on short-term and small spatial scales, and poses an initial obstacle for their receptivity to landscape ecological knowledge. In addition, the abstract nature of landscape ecology and its inherent inability to always provide rapid empirical proof contrasts with customary fields of knowledge such as silviculture. The effects of this unfamiliarity are amplified by the inherent skepticism of practitioners toward a young science and the natural human resistance to change.
- *Unrealistic expectations.* When landscape ecological knowledge is introduced to forest resource managers, most expect to receive prescriptions or ready-made solutions for specific management problems. This expectation leads to disappointment because landscape ecology is more contextual and, in forest management, is used to develop and explore a range of management alternatives rather than to generate specific prescriptions. This situation is compounded when setting of goals is not explicit because forest managers sometimes expect landscape ecological knowledge to generate the missing goals.
- *Viewing GIS technology as a substitute for landscape ecological knowledge.* Although the ready availability of GIS technology and spatial databases assists in the transfer and application of many landscape ecological research findings, the technology may also interfere with transfer and application. Some users involved in policy development, strategic planning, and forest management believe that GIS manipulation of spatial data represents modeling and scientific research; because such explorations do not always include due consideration of the methods, assumptions, logic, or scientific basis for their approaches, the explorations can lead to false premises and entrenchment of misconceptions about patterns and processes in landscape ecology.
- *Information overload.* With the volume of available information increasing so rapidly, users may have access to more scientific knowledge than they can handle, and become overwhelmed. In addition, published scientific knowledge sometimes conflicts, or is duplicated with only subtle differences; as a result, potential users may misunderstand the value and applicability of the available knowledge. The onus is then on the researcher or transfer specialist to discern what knowledge is most relevant or applicable to each user's situation, and to focus on transferring only the most relevant knowledge.

Our experience with these challenges suggests that they are only temporary, though pervasive. Each can be overcome in time with sustained transfer efforts.

Some difficulties may not be readily overcome by transfer efforts alone because the problem lies in organizational cultures, and is more systemic and long term. Nonetheless, it is important for researchers to be aware of these problems, a few of which are outlined below, and to design transfer activities to address them.

- *Audience complexity and diversity.* Landscape-level approaches to forest policy and management often involve an audience hierarchy in which users have different knowledge needs even for the same topic. In addition, various organizations, landowners, and stakeholders are included in policy development, planning, and management. A clear understanding of the roles and responsibilities of these diverse audiences, as well as of their organizational cultures and educational backgrounds, is essential to ensure that each user obtains the knowledge they require in a usable form. This may mean having to transfer similar knowledge in multiple forms to different audiences, which requires flexible approaches and timing.
- *Continuous shifts in organizational priorities.* The need for and use of ecological knowledge by forest managers and policymakers are linked to organizational directions and priorities at any given time. Therefore, any sudden changes in organizational priorities—and these are common and systemic due to social or economic pressures—can also lead to sudden and unexpected shifts in knowledge needs. Adapting knowledge and tools to accommodate such shifts is an ongoing challenge, especially in public agencies.
- *Narrow windows of opportunity for knowledge transfer.* The reality is that most users of forest ecological knowledge, whether they are primarily involved in forest policy development, strategic planning, or forest management, are most receptive to new ecological knowledge when they face a problem and must seek specific solutions under tight time constraints. Although such policy and management crises can provide windows of opportunity for effective transfer, they are narrow and ephemeral. If knowledge developers are not vigilant and do not adapt to such conditions, they may miss many supplementary occasions to transfer knowledge.

In our experience, these challenges are difficult to meet because they require awareness of changing situations and the ability to respond quickly in a manner that is appropriate to each component of the audience. Although knowledge developers and transfer specialists may possess the necessary skills to meet each of these criteria, organizational constraints may prevent them from responding effectively. We are unaware of any general solution to this category of challenges other than to recognize its existence and take measures (e.g., striving to remain aware of the audience's changing context) to detect opportunities sufficiently far in advance to allow an appropriate response.

6.5.2. General Lessons for Landscape Ecologists

Although influencing organizational characteristics to make the situation conducive for successful knowledge transfer is beyond the capacity of landscape ecological researchers and transfer specialists, we believe that several factors are within the realm of their control. The insights we offer below are examples of issues that may be under the direct influence of landscape ecology knowledge developers.

- *More than practitioners can benefit from the transfer of landscape ecological knowledge.* Knowledge developers and transfer specialists can engage a broad range of audiences in addition to forest resource managers, including legislators, policymakers, and land-use planners, who may influence forest management at the many different hierarchical levels involved in solving a forest management problem. Recognizing the specific needs and characteristics of each distinct group of users helps to tailor knowledge transfer efforts accordingly.
- *Knowledge developers need to keep pace with existing policies and practices.* This awareness of the operational context helps researchers and transfer specialists to time the development and transfer of knowledge to match user needs, thereby maximizing effective use and application of the knowledge. When transfer occurs too early, users may be unreceptive because acceptance of the knowledge would demand too big a change from the status quo. If transfer occurs too late, users may no longer need the knowledge (i.e., they may have already developed alternative solutions) or it may no longer be relevant.
- *Continuous engagement and personal interactions are most effective.* Even when users are receptive, continuous engagement by knowledge developers, starting as early as the research design stage, builds mutual trust and facilitates progressive and gradual transfer of knowledge. Continuous engagement also provides opportunities to transfer the same knowledge in different forms to suit different circumstances. As a result, it is a powerful vehicle for knowledge exchange and for increasing the acceptance of new knowledge and its applications.
- *It is essential to establish the context for landscape ecological knowledge at the outset.* This is especially true when knowledge of the underlying concepts must be established before transfer of tools can succeed. Without understanding the concepts, users cannot apply the tools appropriately. Relying on GIS and computing technology supports the transfer of tools in the short term, but may actually impede the transfer of landscape ecological concepts in the long term if those concepts are not made part of the transfer of the tools.
- *A clear understanding of the user's expectations is important for transfer.* In addition to understanding the user's need for specific knowledge or application of the knowledge, researchers must be aware of the user's expectations. Users prefer directly applicable, user-friendly, validated knowledge, whereas researchers may prefer innovative, methodologically elegant, complex solutions to their problems.

In general, we found that the passive approach to knowledge transfer (i.e., expecting users to discover, read, understand, and apply published research knowledge) is ineffective. However, it is possible to provide examples of effective use of supply-driven ("push"), demand-driven ("pull"), and collaborative–iterative modes of active knowledge transfer (Perera et al. 2006) in Ontario. Most early applications of landscape ecology at strategic scales resulted from a push powered by education and the creation of awareness by researchers. This approach was effective in transferring landscape ecological concepts and setting the context at the levels of broad policy development and the production of management guides. Relying on demand (pull) from users continues to be an effective way to transfer landscape ecological tools at the scales of local management and tactical problem-solving, especially once the context is established. Last but not least, the collaborative–iterative approach is optimal in situations such as the development of management guides in which ongoing interaction and adaptability are key to ensuring that the knowledge will be used in an appropriate context and that the tools are adjusted to meet user needs.

6.6. CONCLUSIONS

The case study described in this chapter illustrates that it is possible for a public land management agency to successfully develop and transfer forest landscape ecological knowledge. Policies informed by landscape ecological principles, an awareness of their importance, and an emphasis on implementation have evolved over the past two decades, and practitioners are now using landscape ecological tools to solve forest management problems.

For this to occur, several interrelated enabling factors were essential:

- a combination of political will, driven to some extent by global pressures
- social pressures such as a provincial environmental assessment of forestry practices and new legislation
- enabling structures, including global and local science, provincial policy changes, a supportive organizational structure, and technological advances
- demands for new knowledge and acceptance of this new knowledge by land managers
- adequate resources, including both skilled people and the technological and organizational infrastructure required to support their efforts

All of these factors aligned simultaneously (and fortuitously) and continue to drive the process. Ontario benefited from a combination of these factors along with an institutional capacity for change and flexibility, and a willingness to incorporate new ideas and approaches. The demand for knowledge continues to increase as does the demand for tools to facilitate application of the knowledge by resource managers.

Despite the advantages enjoyed by Ontario, maintaining a connection between the expansion of landscape ecological knowledge and its application in the development of forest policy and in operational practice remains a challenge. Socioeconomic and political realities continue to complicate policy development and management at the broad scales where landscape ecology is most relevant. The structure of Ontario's forest land tenure, with public ownership and private management, and the resulting composition of stakeholders also pose challenges to the application of landscape ecological principles. The adoption process is slowed, for example, by the planning framework (i.e., goal setting) and by long implementation time frames; policy changes made today may not be implemented for up to 10 years, depending on the stage of the forest management planning cycle when the policy changes.

We have learned that just because knowledge is developed, published, and made accessible to practitioners, this does not mean that it will be applied successfully. We recognize that obstacles to successful application will continue to exist and will emerge inevitably at each level, from legislation to policy development, planning, and operational practice. Moreover, we believe that good knowledge transfer is essential, but is only the first step in successful application of landscape ecological knowledge; organizational and other barriers may delay or prevent this application. In this case, a sustained transfer effort is necessary to ensure that the available knowledge will be accepted and used in practice. Success requires dedicated individuals willing to lead, advocate, and push for change over a period of years.

Acknowledgments

We thank Joe Churcher and Jim Baker of the Ontario Ministry of Natural Resources for helpful discussions during the development of this chapter as well as for their thorough reviews of a draft. Fred Swanson of the USDA Forest Service also reviewed a draft and provided much appreciated insight from a perspective outside Ontario.

LITERATURE CITED

AES, CMC, CAI (2000) Review of forest management guides. Consultants report to Provincial Forest Technical Committee. Arborvitae Environmental Services Ltd., CMC Ecological Consulting, and Callaghan & Associates Inc., Toronto.

Band L (1993) A pilot landscape ecological model for forests in central Ontario. Ont Min Nat Resour, Sault Ste. Marie, Ontario. Forest Landscape Ecology Program, Forest Fragmentation and Biodiversity Project Report No. 7.

Brundtland G (ed) (1987) Our common future: The World Commission on Environment and Development. Oxford Univ Press, Oxford, UK.

Certification Canada (2006) Certification status report. Ontario—SFM—December 20 2005. Certification Canada, Ottawa. (http://www.certificationcanada.org/_documents/english/dec2005_sfm_ontario_forest_certification_details.pdf)

Elkie P, Rempel R, Carr A (1999a) Patch Analyst user's manual: a tool for quantifying landscape structure. Ont Min Nat Resour, Northwest Science and Technology, Thunder Bay. NWST Tech Manual TM-002.

Elkie PC, Hooper G, Carr AP, Rempel RS (1999b) Ontario Marten Analyst and Marten1 models: directions for applying the forest management guidelines for the provision of marten habitat in the

Northwest Region. Ont Min Nat Resour, Northwest Science and Technology, Thunder Bay. NWST Tech Manual TM-004.

Elkie PC, Gluck MJ, Rouillard D, Ride K (2002) The natural disturbance pattern emulation guide tool. Ont Min Nat Resour, Northwest Science and Information, Thunder Bay.

Epp AE (2000) Ontario forests and forest policy before the era of sustainable forestry. In: Perera AH, Euler DL, Thompson ID (eds) Ecology of a managed terrestrial landscape: patterns and processes of forest landscapes in Ontario. Univ British Columbia Press, Vancouver, pp 237–275.

Euler DL, Epp AE (2000) A new foundation for Ontario forest policy for the 21st century. In: Perera AH, Euler DL, Thompson ID (eds) Ecology of a managed terrestrial landscape: patterns and processes of forest landscapes in Ontario. Univ British Columbia Press, Vancouver, pp 276–294.

Francis GR (2000) Strategic planning at the landscape level. In: Perera AH, Euler DL, Thompson ID (eds) Ecology of a managed terrestrial landscape: patterns and processes of forest landscapes in Ontario. Univ British Columbia Press, Vancouver, pp 295–302.

Kloss D (2002) Strategic forest management model, version 2 user's guide. Ont Min Nat Resour, Forest Management Planning Section, Sault Ste. Marie.

Li C, Ter-Mikaelian MT, Perera AH (1996) Ontario fire regime model: its background, rationale, development and use. Ont Min Nat Resour, Sault Ste. Marie, Ontario. Forest Landscape Ecology Program, Forest Fragmentation and Biodiversity Project Rep No. 25.

McNicol JG, Baker JA (2004) Emulating natural disturbance: from policy to practical guidance in Ontario. In: Perera AH, Buse LJ, Weber MG (eds) Emulating natural forest landscape disturbances: concepts and applications. Columbia Univ Press, New York, pp 251–264.

Naylor BJ, Baker JA, Hogg DM, McNicol JG, Watt WR (1996) Forest management guidelines for the provision of pileated woodpecker habitat. Version 1.0. Ont Min Nat Resour, Sault Ste. Marie. Tech Series.

Naylor B, Kaminski D, Bridge S, Elkie P, Ferguson D, Lucking G, Watt B (2000) User's guide for OWHAM99 and OWHAMTool (Ver. 4.0). Ont Min Nat Resour, South Central Science Section, North Bay. Tech Rep 54.

OEAB (1994) Reasons for decision and decision. Class Environmental Assessment by the Ministry of Natural Resources for timber management on Crown lands in Ontario (EA-87-02). Ontario Environmental Assessment Board, Toronto.

OMNR (1988) Timber management guidelines for the provision of moose habitat. Ont Min Nat Resour, Wildlife Branch, Toronto.

OMNR (1994) Policy framework for sustainable forests. Ont Min Nat Resour, Sault Ste. Marie, Ontario.

OMNR (1996) Forest management planning manual for Ontario's Crown forests. Ont Min Nat Resour, Sault Ste. Marie.

OMNR (1999) Ontario's living legacy land use strategy. Ont Min Nat Resour, Toronto.

OMNR (2002) Forest management guide for natural disturbance pattern emulation. Version 3.1. Ont Min Nat Resour, Toronto.

OMNR (2003) Old growth policy for Ontario's Crown forests, V.1. Ont Min Nat Resour, Sault Ste. Marie. Forest Policy Series.

OMNR (2004) Forest management planning manual for Ontario's Crown forests. Ont Min Nat Resour, Sault Ste. Marie. Tech Series.

OMNR (2005) Protecting what sustains us: Ontario's biodiversity strategy 2005. Ont Min Nat Resour, Toronto.

Perera AH, Buse LJ, Crow TR (2006) Knowledge transfer in forest landscape ecology: a primer. In: Perera AH, Buse LJ, Crow TR (eds) Forest landscape ecology: transferring knowledge to practice. Springer, New York.

Perera AH, Baldwin DJB, Schnekenburger F (1997) LEAP II: a landscape ecological analysis package for land use planners and managers. Ont Min Nat Resour, Ont For Res Inst, Sault Ste. Marie. For Res Rep 146.

Perera AH, Euler DL, Thompson ID (eds) (2000) Ecology of a managed terrestrial landscape: patterns and processes of forest landscapes in Ontario. Univ British Columbia Press, Vancouver.

Perera AH, Baldwin DJB, Yemshanov DG, Schnekenburger F, Weaver K, Boychuk D (2003) Predicting the potential for old-growth forests by spatial simulation of landscape ageing patterns. For Chron 79(3):621–631.

Perera AH, Yemshanov DG, Schnekenburger F, Baldwin DJB, Boychuk D, Weaver K (2004) Spatial simulation of broad-scale fire regimes as a tool for emulating natural forest disturbance. In: Perera AH, Buse LJ, Weber MG (eds) Emulating natural forest landscape disturbances: concepts and applications. Columbia Univ Press, New York, pp 112–122.

Racey G, Harris A, Garrish L, Armstrong T, McNicol J, Baker J (1999) Forest management guidelines for the conservation of woodland caribou—a landscape approach. Version 1.0. Ont Min Nat Resour, Thunder Bay.

Schnekenburger F, Perera AH (2003) Regional-scale forest net primary productivity software (NPPAS): a user's guide. Ont Min Nat Resour, Ont For Res Inst, Sault Ste. Marie, Ontario. For Res Inf Pap 156.

SPS (2006) Patchworks user guide. Spatial Planning Systems, Deep River, Ontario. (www.spatial.ca)

Statutes of Ontario (1995) Crown Forest Sustainability Act, revised. R.S.O. 1998. Chapter 25 and Ontario Regulation 167/95.

Voigt DR, Broadfoot JD, Baker JA (1997) Forest management guidelines for the provision of white-tailed deer habitat. Version 1.0. Ont Min Nat Resour, Sault Ste. Marie. Tech Series.

Watt WR, Baker JA, Hogg DM, McNicol JG, Naylor BJ (1996) Forest management guidelines for the provision of marten habitat. Version 1.0. Ont Min Nat Resour, Sault Ste. Marie. Tech Series.

7

Moving to the Big Picture: Applying Knowledge from Landscape Ecology to Managing U.S. National Forests

Thomas R. Crow

7.1. Introduction
7.2. National Forest Planning
    7.2.1. Background
    7.2.2. Examples
    7.2.3. Challenges
7.3. Regional and National Resource Assessments
    7.3.1. Background
    7.3.2. Examples
    7.3.3. Challenges
7.4. Analyses of Landscape and Regional Change
    7.4.1. Background
    7.4.2. Examples
    7.4.3. Challenges
7.5. Integrated Landscape Management
7.6. Emulating Natural Disturbance
7.7. Managing Roads
7.8. Conclusions
    Literature Cited

THOMAS R. CROW • USDA Forest Service, Research and Development, Environmental Sciences, 1601 N. Kent Street, Arlington, VA 22209, USA.

7.1. INTRODUCTION

United States National Forests encompass 77.7 million ha (192 million acres) of grasslands and forests, which comprise 7% of the nation's total land base and 20% of the nation's forested lands. Increasing demand for wood has raised concerns about producing forest products without impeding the land's ability to provide a variety of other renewable goods and ecosystem services (Aber et al. 2000). Land-use conflicts often arise that result in challenges to forest plans and, in many cases, costly and time-consuming litigation. A more comprehensive planning and management approach is needed that allows public lands to generate multiple values and benefits. Landscape ecologists are among those contributing concepts, perspectives, and information to help meet this need (e.g., Forman 1995; Lindenmayer and Franklin 2002; Liu and Taylor 2002; Wiens and Moss 2005).

The science of landscape ecology is applied to planning and managing the U.S. National Forests in at least six broad areas: National Forest planning, regional and national resource assessments, analyses of landscape and regional change, integrated landscape management, emulating natural disturbance in forest management, and managing roads. In this chapter, I explore the transfer of knowledge and technology from the science of landscape ecology to National Forest planners and managers in each of these topic areas. Because National Forests are part of broader landscapes with multiple ownerships, I do not focus solely on federal lands in this chapter. Indeed, a critical question concerns the role that public lands play and the unique opportunities they provide within the broader landscape context, which is characterized by multiple ownerships and varied management objectives.

To be consistent with other chapters in this book, I distinguish among technology transfer (tools, data, models), knowledge transfer (concepts and principles), and the process of transferring or communicating these tools and concepts. The transfer of knowledge and technology from science into practice ranges from informal and individual to formal events with broad participation and a national scope. In some cases, old but proven technologies have been utilized, such as revised timber management guides that include a landscape perspective (e.g., Gilmore et al. 2004); in other cases, new technologies are being employed, such as computer visualization (e.g., Wang et al. 2006), succession and disturbance simulations (e.g., Chew et al. 2004; Keane et al. 2002), or Web-based interactive models (e.g., HARVEST LITE; Gustafson and Rasmussen 2002). Throughout this chapter, examples of approaches are presented for transferring knowledge and technology into practice.

7.2. NATIONAL FOREST PLANNING

7.2.1. Background

The National Forest Management Act of 1976 commits the USDA Forest Service to managing National Forest lands according to land and resource management plans that provide for multiple uses and sustained yield of renewable resources. Currently, these plans—commonly called "forest plans"—are being revised. A "landscape perspective" is evident in these revisions. Specific topics in which concepts from landscape ecology are currently contributing to planning include:

- practicing stewardship across ownership boundaries
- using ecosystems as fundamental management and planning units
- allocating multiple uses in time and space
- managing landscape composition and structure to meet diverse management goals
- quantifying the cumulative impacts of local practices at larger spatial and temporal scales
- planning and managing at multiple spatial and temporal scales
- considering the social, economic, and ecological contexts in forest planning and implementation of the plans

7.2.2. Examples

Many issues common to forest planning—including management of old-growth forests, protection of threatened and endangered species, preservation of forest health, prevention of wildland fire, and wilderness management—necessitate broad-scale approaches to resource management. The Northwest Forest Plan, for example, addresses management on federal lands (including USDI Bureau of Land Management and National Parks land, and USDA Forest Service National Forests) across 9.7 million ha (24 million acres) in three states—Oregon, Washington, and northern California, defined primarily by the range of the northern spotted owl (*Strix occidentalis caurina*)—where federal lands are designated as either protected reserves or matrix lands that can be harvested (FEMAT 1993). The Plan is an early attempt at a comprehensive, ecosystem-based approach to public land management—that is, managing whole systems, including local, landscape, and regional ecosystems, and broad assemblages of plants and animals—meshed with the more common emphasis on individual forest stands and individual species, such as the northern spotted owl and the marbled murrelet (*Brachyramphus marmoratus*; Diaz 2004, FEMAT 1993). I consider the Northwest Forest Plan along with other regional and national resource assessments in more detail later in this chapter.

A landscape perspective is also apparent in the Chief of the USDA Forest Service's list of perceived threats to the nation's forests. Among these threats is the loss of open spaces, which includes fragmentation caused by land development and especially by the urbanization of private lands within and near public forests. Increasingly, National Forests are becoming islands of wild and semiwild land embedded within a matrix of developed lands. Agency managers recognize that landscape change outside the boundaries of National Forests has important implications for management within their boundaries.

Concepts and principles from landscape ecology are helping managers to address other perceived threats as well—forest health threats, wildland fire, invasive species, and use of off-road vehicles, among others. Each of these threats requires approaches that allow managers and planners to consider spatial relationships. For example, the Healthy Forests Restoration Act of 2003 (P.L. 148-108; http://www.healthyforests.gov/initiative/legislation.html) directs the USDA Forest Service and USDI Bureau of Land Management to plan and conduct projects to reduce hazardous accumulation of fuels so as to reduce the risk from wildfire and to improve forest and rangeland health. A critical question related to implementing this Act concerns a spatial element: where in the landscape should fuel reduction treatments be applied to maximize their benefits? Research conducted on predicting forest fire behavior and effects at the landscape level (e.g., Finney 1999; Gardner et al. 1999) has helped to address this question, but efforts to date have failed to provide managers with the tools necessary to more fully consider the various trade-offs when altering the composition, structure, and function of landscapes for a single purpose—to defuse the fire bomb.

Furthermore, there are questions related to assessing the effectiveness of fuel-reduction treatments. By necessity, treatments are local in their application, but there is increasing recognition that factors operating at the regional, subcontinental, and even continental scales are shaping local conditions (Hansen et al. 2001; Neilson 1995; Swetnam and Betancourt 1990). An important lesson learned from addressing the Chief's four threats, including forest health and fires, is the need to manage natural resources at multiple spatial and temporal scales. None of these threats can be resolved at a local scale alone nor can any of the threats be resolved independently of other important natural resource issues that are often regional or national in scope.

A spatial framework for management treatments, including fuel-reduction treatments, is a precursor for an ecosystem-based approach to land management (Crow 2002). Spatially explicit landscape models provide a means for adding this framework. Most spatial models, however, are designed as research tools and relatively few are available with "off the shelf" capabilities for management and planning. One such model is SIMPPLLE, the acronym for Simulating Patterns and Processes at Landscape Scales, which was designed primarily as a management and planning tool to formally incorporate spatial considerations into designing and evaluating land management alternatives over a range of spatial scales (Chew et al. 2004). The model is designed to:

- use existing inventory data, where possible, as the input in a polygon or grid format, with ArcView and ArcGIS extensions providing spatial outputs
- treat disturbances as probabilistic events
- distribute disturbance spatially within the landscape
- quantify the range of variability for vegetation conditions and disturbance processes
- simulate interactions among disturbances and vegetation patterns
- project future conditions under a variety of management options
- integrate knowledge from research with expert opinion

Terrestrial and aquatic ecosystems are recognized in the model, with terrestrial ecosystems including both forests and grasslands. Linkages between SIMPPLLE and scheduling and optimization models such as MAGIS and SPECTRUM can aid in evaluating alternative management scenarios (Zuuring et al. 1995). For example, SIMPPLLE can be used to assess health risks within the landscape based on the interactions among multiple stressors, then MAGIS can be used to schedule management activities to reduce the perceived risk to forests. The application of SIMPPLLE is required as part of management plan revision by the USDA Forest Service and the USDI Bureau of Land Management in Montana and Idaho (J.D. Chew, USDA Forest Service, Rocky Mountain Research Station, personal communication).

The timber harvest allocation model HARVEST is another example of a landscape model designed for practical application (Gustafson and Crow 1996). HARVEST provides a visual and quantitative means for predicting the spatial patterns produced by even-aged harvesting strategies. Timber harvests are allocated using a digital stand map in which the values for each cell in the grid represent stand age. The modeler specifies the size distribution of the harvests, the total area of forest to be harvested, the rotation length, and the width of buffers between adjacent harvested areas. HARVEST has been used to project landscape patterns under alternative forest plans for the Hoosier National Forest in Indiana (Gustafson and Crow 1996). The initial forest plan called primarily for clearcutting to be distributed across most of the National Forest; an amended plan featured group selection (harvesting in small groups) across a limited portion of the National Forest. HARVEST provided the means for projecting the landscape patterns produced under these two management scenarios over several timber rotations. As expected, these scenarios created two very different landscapes in terms of patch-size distribution and the amount of forest edge and forest interior that is present.

7.2.3. Challenges

Each application of a management or planning model offers an opportunity to transfer technology (e.g., Gustafson et al. 2006) and knowledge (e.g., Lytle et al. 2006) into practice. This transfer takes place in a variety of venues, from formal training sessions for managers conducted by modelers to joint projects involving researchers and managers in applying models such as SIMPPLLE and HARVEST to forest management planning. A Web-based model, HARVEST LITE (Gustafson and Rasmussen 2002), is now available that allows users to easily compare alternative harvesting strategies by changing harvest size, spatial distribution, and intensity (expressed as the area harvested per decade) within the limited range of simulated landscapes. By using these landscapes and limiting the amount of model parameterization, users can easily evaluate and compare a large number of management scenarios. This hands-on approach has been especially effective in workshops and training sessions for managers, where the model becomes the means for visualizing the outcomes of management decisions at a landscape level.

Successful implementation of the Healthy Forests Restoration Act of 2003 requires the USDA Forest Service to implement a major effort in technology and knowledge transfer. In support of these national-level transfer efforts, practical guidelines are being developed and presented as "desk guides" for managers. As with all national efforts, guidelines must be flexible enough to allow for differences in local conditions and sufficiently detailed to provide useful guidelines for application in the field. Meeting these standards is one of the main challenges in transferring knowledge into practice.

7.3. REGIONAL AND NATIONAL RESOURCE ASSESSMENTS

7.3.1. Background

The USDA Forest Service and other federal and state agencies have conducted numerous broad-scale biophysical and social assessments (Table 7.1) in response to a variety of issues and needs (Jensen and Bourgeron 2001; Johnson et al. 1999). In doing so, planners and managers are thinking beyond the boundaries of their National Forests, and taking into account the social, economic, and ecological contexts in which they manage public lands. The issues that provide the catalyst for regional assessments include fire danger, forest health, endangered species, and old-growth forests in assessments such as the Sierra Nevada Ecosystem Assessment in California and the Interior Columbia Basin Ecosystem Management Project (ICBEMP) in eastern Oregon and Washington, Idaho, and western Montana. The issues can also expand to include the need for considering more integrated management strategies at the stand and landscape levels, as is the case in the Ozark–Ouachita Highlands Assessment in Arkansas, Missouri, and Oklahoma, and the Great Lakes Ecological Assessment in Minnesota, Wisconsin, and Michigan (Table 7.1). Not all assessments had specific statutory mandates, but common to all assessments was a need to consider the broader landscape and regional conditions, trends, and resource issues in order to adequately plan within the boundaries of the National Forests (GAO 2000). Four of the assessments—the Northwest Forest Plan, ICBEMP, the Northern Forest Lands Study, and the Southern Forest Resources Assessment—are profiled below.

7.3.2. Examples

The Northwest Forest Plan remains one of the boldest efforts undertaken by a federal agency to implement adaptive management at the landscape and regional levels. After 10 years of this experiment in landscape and regional management on the western side of the Cascades, there has been little harvesting in either reserve or matrix lands, and as a result, Moeur et al. (2005) estimated a net increase of 251 000 ha (620 000 acres) of forest with trees greater than 51 cm (20 inches) in diameter at breast height (DBH). Despite this trend, the population of spotted owls declined on average by 3.7% per year from 1990 to 2003 (Lint 2005). Furthermore, while forests on federal lands are maturing, forests on private lands often are being intensively managed for timber, producing a contrast in forest structure between private and public lands.

The ICBEMP, initiated in 1994 and concluded in 2003, was a large, multiowner, interdisciplinary project encompassing 58.7 million ha (145 million acres) and 64 different jurisdictions, with an integrated terrestrial and aquatic assessment. The plan was implemented for public lands managed primarily by the USDA Forest Service and the USDI Bureau of Land Management (Quigley and Arbelbide 1997; Quigley et al. 1996). The issues driving the ICBEMP broadly relate to forest health and include the threats of wildfire and invasive species, as well as the protection and restoration of habitat for fish and wildlife species. Current landscapes within the interior basin are at greater risk of fire, insect infestation, and disease than under historical conditions (Hessburg et al. 1999), rangelands are highly susceptible to invasive species (Bunting et al. 2005), and aquatic systems are more fragmented and isolated than was historically the case, and are vulnerable to the introduction of nonnative fish species that threaten native species (Rieman et al. 2000). Unlike the Pacific Northwest Plan, however, the ICBEMP did not result in a formal regional plan for managing public lands; instead, the decision was made to incorporate the research findings into ongoing USDA Forest Service planning efforts. This piecemeal approach to applying the results from ICBEMP produced uneven applications at best and, in many respects, this decision negated many of the advantages offered by the landscape and regional perspective.

In 1988, Congress directed the USDA Forest Service to cooperate with several States in the Northern Forest Lands Study, which examined the timberland resources in northern New York, Vermont, New Hampshire, and Maine in order to assess the current condition of the forest resources and to develop alternative strategies that would protect the long-term integrity and traditional uses of the land (Harper et al. 1990). There were concerns about the future of the 10.5 million ha (26 million acres) of mostly private forest land in this four-state area of the United States. Changes in land ownership—specifically, the fragmentation of ownership in which large blocks of private forested lands were being subdivided into smaller parcels and, in many cases, developed to provide second homes and other residential uses—threatened the long-term integrity and traditional land uses in many parts of New England and northern New York. Within the Northern Forest Lands area, land adjacent to lakes and rivers and land with a scenic vista (such as ridge tops) are the most vulnerable to changes in land use. Proximity to highways and secondary roads also increases the likelihood of development. In their final report to the U.S. Congress and State Governors, the Task Force responsible for the study identified the important natural resources of the region, and established priorities and guidelines for conserving these resources at the landscape and regional levels (Harper et al. 1990). Twenty-eight conservation strategies were proposed for six broad areas: using land-use controls and planning for conservation, using easements and land purchases to meet conservation goals, maintaining large contiguous tracts of forest ownership by providing incentives to not fragment the land, combining community

Table 7.1. A summary of recent regional assessments conducted by the USDA Forest Service and their partners

Regional assessments	Area	Partners[a]	Mission
Forest Ecosystem Management Assessment (Northwest Forest Plan)	West of the Cascades divide in Washington, Oregon, and northern California	USFS, NOAA, National Marine Fisheries, BLM, F&W, NPS, and EPA	Define an ecosystem-management approach to sustain biological diversity, maintain long-term site productivity, and sustain natural resources, including timber.
Interior Columbia Basin Ecosystem Management Plan (ICBEMP)	58.7 million ha (145 million acres) of the Columbia Basin east of the Cascades crest	USFS, BLM, EPA, NOAA, and F&W	Provide the scientific basis for managing public lands in the Interior Columbia Basin to meet community needs in an ecologically sustainable way.
Southern Appalachian Assessment	15 million ha (37 million acres) from West Virginia and northeastern Virginia to northwestern South Carolina, northern Georgia, and northern Alabama	Southern Appalachian Man and Biosphere (SAMAB) Program	Summarize what is known about the regional ecosystems (their air, water, land, and people), and identify current and emerging resource-management problems.
Great Lakes Ecological Assessment	20.6 million ha (51 million acres) in Minnesota, Wisconsin, and Michigan	USFS, EPA, NRCS, NBS, States, and universities	Define scope and context for major resource management issues. Provide information about current status of regional forests to promote collaborative planning.
Sierra Nevada Ecosystems Project (SNEP)	The entire Sierra region of California and Nevada, including 11 National Forests (40% of the area)	Independent panel (i.e., non-USFS) of scientists	Assess the health and sustainability of the Sierra Nevada forest ecosystems. Provide strategies for protecting the health and sustainability of the forest while providing for human needs.

Northern Forest Lands Study	10.5 million ha (26 million acres) in Maine, northern Vermont and New Hampshire, and northeastern New York	States of Maine, Vermont, New Hampshire, and New York	Document recent changes in land ownership and land use that can be used for developing a common vision for the future of the regional forests.
Southern Forest Resource Assessment	Virginia, Kentucky, Tennessee, North Carolina, South Carolina, Florida, Alabama, Mississippi, Louisiana, Arkansas, and eastern Texas	EPA, USGS, ACE, NPS, NBS, TVA, ORNL, F&W and States	Provide information about the current status and project likely future conditions of the regional forest in the South to enhance planning and management of the resource.
Ozark–Ouachita Highlands Assessment	107 counties, including 2.6 million ha (6.5 million acres) of state and federal lands in Arkansas, eastern Oklahoma, and southern Missouri	USFS: Eastern Region, Southern Region, North Central Research Station, and Southern Research Station	Characterize current status and trends within the study area for social and economic conditions; aquatic and terrestrial vegetation and wildlife; and air quality.

[a] ACE, Army Corps of Engineers; BLM, Department of the Interior, Bureau of Land Management; EPA, Environmental Protection Agency; F&W, Fish and Wildlife Service; NBS, National Biological Service; NRCS, U.S. Department of Agriculture, Natural Resources Conservation Service; NOAA, National Oceanic and Atmospheric Administration; NPS, National Parks Service; ORNL, U.S. Department of Energy, Oak Ridge National Lab; TVA, Tennessee Valley Authority; USFS, U.S. Department of Agriculture, Forest Service; USGS, U.S. Geological Service.

development with land conservation, keeping private land accessible to the public, and developing coordinated regional plans.

The recommendations of the Northern Forest Lands Plan were presented to nearly 1000 people who attended 21 public meetings (Harper et al. 1990). A consistent message throughout the study was the need for greater coordination and cooperation between public and private forest owners in planning and management at the landscape and regional levels. The goal of this effort was not to create new regulations, but rather to inform stakeholders and create the public awareness that is the prerequisite for political action.

The Southern Forest Resource Assessment was initiated in 1999 because of an expressed desire by natural resource managers, scientists, and the public to better understand current conditions as well as the forces shaping the future forest in the South (Wear and Greis 2002, 2003). Thus, the Southern Forest Assessment was not conducted in response to an immediate crisis or conflict, but rather to address long-term concerns about the effects of rapid urbanization of forested land, increasing demand for timber, declining forest health, and increasing air pollution on the future of the region's forests. As with most regional assessments, federal, state, and local partners participated (Table 7.1). Among the 81.3 million ha (201 million acres) of commercial forested land in the South, 89% is privately owned (Wear and Greis 2002). Ownership by timber companies has decreased during the past several decades, while ownership by investment companies has increased.

During the past 25 years, both timber harvesting and urbanization of timberlands have increased dramatically in the South, but neither can continue to increase indefinitely. Furthermore, invasive species, including diseases and insects, are having a significant impact on the health of southern forest ecosystems. Urbanization could also increase these impacts (Wear and Greis 2002). An important finding drawn from the assessment is the conclusion that "urbanization presents a substantial threat to the extent, condition, and health of forests." Among the forces of change in forested land, urbanization will have "the most direct, immediate, and permanent effect" at the landscape and regional levels (Wear and Greis 2003, p. 92).

A periodic national assessment of forests in the United States is required by the Forest and Rangeland Renewable Resources Planning Act (RPA) of 1974. These periodic surveys provide information about the current status of the nation's forests as well as trends in their condition. The most recent national assessment, in 2002, was the fourth national assessment to be conducted, and covered a variety of topics. These included: conserving biological diversity, maintaining the productive capacity of forest and rangeland ecosystems, maintaining forest health and vitality, contributing to carbon sequestration, meeting the needs of society, and the legal, institutional, and economic frameworks for conserving and sustaining forests (USDA Forest Service 2001). RPA reports provide information about historical and projected supply and demand for timber at regional and national scales. For years, these reports have been effective in shaping perceptions about future commodity demands and supplies at these spatial scales. These perceptions, in turn, help guide forest policy in National Forests, and for that matter, in all forest ownerships in the United States.

Regional assessments provide an ideal perspective for identifying ecosystems at risk. In the Southern Forest Assessment, a total of 14 critically endangered forest ecosystems were listed as having been greatly reduced in their extent since European settlement. Among these are old forests of all types, high-elevation spruce–fir (*Picea–Abies*) forests, a variety of wetlands, bog complexes and pocosins (bogs that form in shallow, nondraining depressions) throughout the South, bottomland and flood-plain forests, open lands (including glades, barrens, and prairies), and longleaf pine (*Pinus palustris* Mill.) forests and Atlantic white cedar (*Chamaecyparis thyoides* L.) swamps (Wear and Greis 2002). Given the ownership patterns in the South, most of these at-risk ecosystems (with the exception of old forests and high-elevation spruce–fir) occur on private land, so conservation strategies necessitate the involvement of multiple owners.

Landscape ecologists stress the importance of spatial context when evaluating local management opportunities. Although public ownership, including National Forests, represents only a small portion of the South's commercial forests, it provides unique ecological and social values within the region. When viewed at a regional level, public lands provide much of the interior (nonedge) forest habitat and a disproportional amount of the mature forests in the South. These represent both opportunities and responsibilities for public land managers.

7.3.3. Challenges

Regional assessments are an essential part of the National Forest planning process. They provide critical information for making local decisions and for setting the management direction for obtaining the desired future conditions within a National Forest's boundaries. There is, however, no clear legal mandate to conduct these assessments, funding to conduct regional assessments is often limited, and National Forest supervisors are not obligated to formally incorporate regional findings into their forest planning. In a recent study of USDA Forest Service planning and the Great Lakes Ecological Assessment (GAO 2000), the United States General Accounting Office (GAO) concluded that better integration of broad-scale assessments is needed for National Forest planning. The GAO report makes a number of useful recommendations to maximize the value of broad-scale biophysical and social assessments in forest planning (Table 7.2).

Conveying the information contained in these regional assessments to a variety of audiences, from professional land managers to the general public, is an ongoing challenge. In most cases, technical reports are published and then findings are presented in public forums and in newspaper articles in order to make the results available and hopefully meaningful to the general public. In many cases, the land management issues are sufficiently contentious that press coverage is substantial but not necessarily informative. The challenge, as always, is to present complex issues in a straightforward and understandable way.

For natural resource managers and planners, the story is more encouraging. Publications such as Jensen and Bourgeron's (2001) *A Guidebook for Integrated*

Table 7.2. United States General Accounting Office (GAO 2000) suggestions for increasing the value of regional assessments in National Forest planning

Guidelines for conducting regional assessments
- Assessments should occur early in the planning process.
- The process of conducting an assessment should be open to all interested parties.
- Clear objectives and identifiable products are needed prior to conducting the assessment.
- The geographic scope of the assessment should coincide with the nature of the issues to be addressed.
- To be effective, both federal and nonfederal lands need to be included in the assessment.
- Assessments include gathering information, analyses, and conclusions, but do not include making decisions.
- Realistic estimates of costs for conducting assessment are essential.
- Secure funding, specifically for the purposes of conducting assessments and reporting the results, is essential.
- Support for regional assessments is needed at the highest levels of the organization.

Ecological Assessments and Johnson and colleagues' (1999) *Bioregional Assessments* provide useful guidelines for conducting regional ecological assessments. In both cases, the guidelines are based on the practical experience of conducting regional assessments, and the authors use case studies (e.g., Great Lakes Ecological Assessment, Northern Forest Lands, Southern Appalachian Assessment, Upper Mississippi River Adaptive Environmental Assessment) to share their experiences with professional managers and planners. Publication in professional journals is another means for transferring knowledge about regional assessments. Wear and Greis (2002), for example, provide a useful summary of the Southern Forest Resource Assessment in a *Journal of Forestry* paper, and Haynes et al. (1998) used the same journal to explore the relationship between science and management based on their ICBEMP experience.

Active programs of technology and knowledge transfer are common to regional assessments. Most assessments have technology transfer or communication plans in place; however, a formal mechanism for evaluating the effectiveness of these transfer efforts is generally lacking.

7.4. ANALYSES OF LANDSCAPE AND REGIONAL CHANGE

7.4.1. Background

An expanding program of research within the USDA Forest Service, conducted in cooperation with university researchers, is aimed at better understanding the complex interactions among changes in landscape composition and structure, the factors driving change, and the ecological, social, and economic implications of the change. A number of interrelated landscape issues—including urban sprawl, forest fragmentation, forest health, loss of open spaces, invasive species, and forest productivity—are relevant to National Forest managers.

Many factors contribute to landscape change. Schulte et al. (2003) studied changes in the composition and age structure of regional forests in the Lake States that reflected both natural and human-related causes. In Michigan, for example, the aspen–birch (*Populus–Betula*) type has decreased by nearly 0.8 million ha (2 million acres) since 1935, while during the same period, the maple–beech–birch (*Acer–Fagus–Betula*) type increased by almost 1.0 million ha (2.5 million acres). These compositional changes have implications for forest productivity and carbon sequestration. As the fast-growing aspen becomes less abundant and the slow-growing maple becomes more prevalent in the regional forest, declines in regional forest productivity are likely to occur even with significant investments in silvicultural treatments. In the Lake States, conifers such as hemlock (*Tsuga canadensis* [L.] Carr.), white pine (*Pinus strobus* L.), red pine (*P. resinosa* Ait.), and jack pine (*P. banksiana* Lamb.) have declined in abundance since the original land survey (Schulte et al. 2003). In Figure 7.1, this information on changes in the dominance of conifers is plotted for ecological units, represented in this case by regional ecosystems or sections embedded within a Province of the Great Lakes region (Albert 1995), thus providing a means for displaying large amounts of geographic information in a concise way.

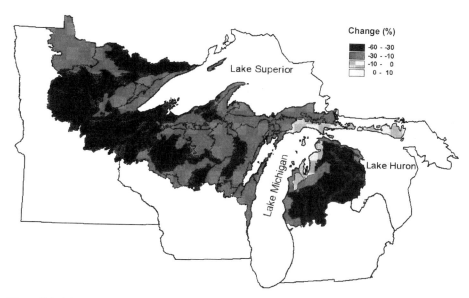

Figure 7.1. The change in the relative dominance (%) by conifers from presettlement times to the present for the regional ecosystems (sections) within Province 212 of the Great Lakes Region (Albert 1995). Presettlement values for relative dominance were based on Government Land Survey records (see Schulte et al. 2003; Schulte and Mladenoff 2001). Present values are based on Forest Inventory and Assessment (FIA) measurements.

7.4.2. Examples

The increased area of loblolly pine (*Pinus taeda* L.) plantations represents a significant change in regional forest composition. Before 1950, less than 1 million ha (2.5 million acres) of pines had been planted in the South; now there are more than 10 million ha (24.7 million acres) (Alig and Butler 2004). Higher timber prices are projected to result from the conversion of about 3 million ha (7.4 million acres) of forested land into agricultural land during the next 30 years unless there is an offsetting flow of land from agricultural production into forest cover (Alig and Butler 2004). The shifts in forest cover noted by Schulte et al. (2003) and Alig and Butler (2004) could significantly reduce regional carbon stores (Emanuel et al. 1984; Pennock and van Kessel 1997).

Changes in the connectivity of the forest landscape are also apparent. Fragmentation indices measure the extent to which patches of forest habitat have been subdivided and dispersed. Forest fragmentation is routinely influenced by human activities and is especially pervasive as a result of urbanization, agricultural activities, and timber harvesting (Riitters et al. 2000, 2002; Wade et al. 2003). Based on the analysis of land-cover maps with a 30-m resolution for the conterminous United States, Riitters et al. (2002) found that overall, 43% of the nation's forests were located within 90 m of a forest edge and 62% were located within 150 m of an edge. They concluded that this fragmentation is so pervasive that edges affect ecological processes in almost all forested land in the United States.

As mentioned in the previous section on regional and national resource assessments, a common source of forest fragmentation is an increase in the number of owners or the subdivision of larger landholdings into smaller blocks. Concerns about this process were a primary reason for conducting the Northern Lands Study and the Southern Forest Assessment (Table 7.1). The overriding concern is that this fragmentation will result in urbanization and conversion of forest and other open lands into other built-up land uses (Gobster and Rickenbach 2004). Another concern is that smaller parcels may not be economically viable for timber production (Mehmood and Zhang 2001). Although fragmentation has been occurring for a long time in the United States, the rate and extent have increased dramatically in recent years, due in large part to what Hammer et al. (2004) call the "spatial deconcentration" of human populations during the twentieth century and the associated expansion of human settlements (Figure 7.2). The net result is that small increases in human population can cause very large changes in the composition and structure of the landscape. Understanding where people choose to live provides valuable insights about the factors that drive landscape change (Dwyer and Childs 2004; Stewart et al. 2004).

7.4.3. Challenges

A great deal of information regarding landscape change has been effectively conveyed to broad audiences using Web sites. Figure 7.2, for example, is available on

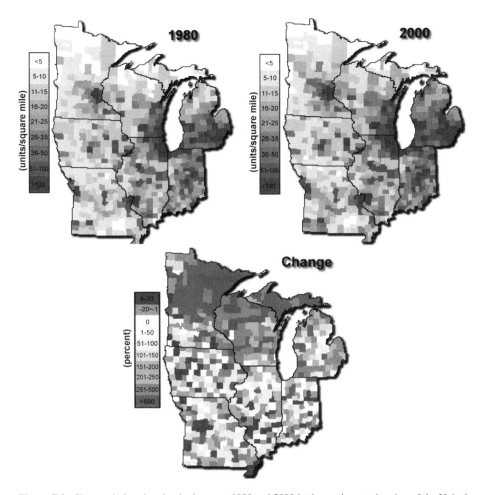

Figure 7.2. Changes in housing density between 1980 and 2000 in the north-central region of the United States. Major increases in housing density in the upper part of the region have occurred without major increases in human population because many new sites represent second homes for people already living in major metropolitan areas. Source: The Changing Midwest Assessment (http://ncrs.fs.fed.us/4153/deltawest/) (*See Colour Plates between pages 132–133.*).

the Web site for "The Changing Midwest Assessment," which is maintained by the Landscape Ecology Research Work Unit of the USDA Forest Service's North Central Research Station (http://ncrs.fs.fed.us/4153/deltawest/). Broad regional trends are readily apparent when change maps such as the example in Figure 7.2 are created. In addition to the trends, however, the implications of these trends must be articulated in terms that make sense to the public.

7.5. INTEGRATED LANDSCAPE MANAGEMENT

The term "landscape management" is now commonly used in the USDA Forest Service. The adoption of this terminology reflects the desire to improve stewardship across ownership boundaries, to better assess the cumulative impacts of many local decisions, to better understand the interactions between land and water, and to develop a more spatially defined approach to resource management. Often, the focus is on a specific geographic area such as the Greater Yellowstone Ecosystem, the Oregon Coast Range, the Chesapeake Bay Watershed, or the southern Appalachians.

The Coastal Landscape Analysis and Modeling Study (CLAMS), a large interdisciplinary effort being conducted by scientists at the USDA Forest Service's Pacific Northwest Research Station and university cooperators, supports a more holistic approach to resource management (Spies 1998). The CLAMS study area includes more than 2.0 million ha (5 million acres) of mixed ownership and is designed to help managers and planners evaluate the aggregate effects of different forest policies and practices on the ecological and socioeconomic conditions within the study area. Using both field and satellite information, researchers produce maps of current vegetation and use models to project changes in vegetation, wildlife habitat, and land use through time (Spies et al. 1994). This is a pioneering "big picture" approach to resource management across many ownerships over a large area in which federal lands such as the Siuslaw National Forest in the Oregon coastal range are only one part of the total picture. Similar studies are occurring elsewhere in the United States (e.g., the Lower Mississippi Alluvial Valley, the Chesapeake Bay Watershed) as managers give greater attention to stewardship across ownership boundaries.

Spatially explicit models of landscape dynamics are essential tools for integrated landscape management. Computer-generated animation that is developed as the output from such spatial models is especially useful for evaluating management scenarios at both stand and landscape scales (McGaughey 1998; Muhar 2001). Visualization tools have been linked to forest growth simulators, and a three-dimensional "flyover" of real landscapes is possible by "draping" GIS maps over digital elevation models (Wang et al. 2006). Although still in development, these technologies offer great potential for applying integrated landscape management and for transferring knowledge of landscape ecology. A variety of audiences, from professional land managers to the general public, can use realistic animations to simulate or understand the effects of management within a landscape. When the landscape being considered is their "home place," interest is especially high and opportunities for meaningful public participation in deciding the desired future conditions within the landscape are greatly enhanced.

The Minnesota Forest Resources Council is leading a successful effort in integrated resource management in which a wide range of interests—for example, commercial logging contractors, representatives from labor organizations, environmental interests, nonindustrial private forest landowners, tribal representatives, and State and federal agencies—are working together to delineate regional landscapes within the State, to identify principles and goals that help guide landscape-based planning and coordination, and to establish a general landscape-based planning process (see http://www.frc.state.mn.us/Landscp/Landscape.html for more information). Planning is accomplished by a volunteer, citizen-based "regional landscape committee" for each of eight regional landscapes within Minnesota. Fundamentally, the process is one of building trust and building relationships. Without these two prerequisites, there can be no landscape-based planning and coordination. Even when these conditions are present, the process can be messy. Partners can drag their feet on decisions, can passively resist, and can leave the table. This is inevitable given the imperfect nature of practicing stewardship across ownership boundaries.

7.6. EMULATING NATURAL DISTURBANCE

There is a growing interest in emulating natural disturbance and using knowledge of the landscape dynamics associated with natural disturbances as a guide for conducting management practices in National Forests in the United States (e.g., Swanson et al. 1997, Wallin et al. 1996, Wimberly et al. 2004; Zasada et al. 2004). The underlying assumption is that forest ecosystems have intrinsic properties that are related to the frequency, duration, and intensity of disturbance. If management impacts fall within the range of variability defined by historical natural disturbance, it is thought that the managed forest ecosystems are more likely to be sustainable (Landres et al. 1999). Thus, emulating natural disturbance has emerged as a means for achieving forest sustainability (Perera et al. 2004).

The general concepts that define this approach have taken several forms, including silvicultural applications (Bergeron and Harvey 1997; McRae et al. 2001), disturbance and forest dynamics (Armstrong 1999; He et al. 2004a), decision-support systems (Hessburg et al. 2004), and forest harvesting patterns (Franklin and Forman 1987; Gustafson and Crow 1996; Li et al. 1993). Landscape ecologists have made significant contributions to these topics.

The Augusta Creek Study, conducted in the Willamette National Forest in western Oregon, is a good example of applying the concept of emulating natural disturbance in the field. Here, a spatially and temporally explicit landscape plan was developed for a 7600-ha area (18 780 acres) with the primary objectives of maintaining native species, ecosystem processes, and landscape structures, and of maintaining long-term ecosystem productivity in a landscape where much of the area is allocated to timber management (Cissel et al. 1998). Although this intermediate step is a common operational step in the forest planning process, there are three aspects that make the Augusta Creek Study a useful guide for others.

First, historical fire regimes are used as the basis for vegetation management. Past fire frequencies, intensities, and spatial patterns were used as a template to guide rotation lengths, harvest rates, green-tree retention levels, and the spatial pattern of timber harvests. As in all such applications, the underlying assumption is that native species are adapted to the range of patterns created by historical disturbances. A second feature of the Augusta Creek Study is the integration of terrestrial and aquatic management objectives through the use of a landscape perspective. Specifically, the management of aquatic ecosystems was designed to be complemented by upslope management practices and patterns given both the larger landscape prescriptions and local conditions (Cissel et al. 1998). As this suggests, a third element was the linkage of management objectives across spatial scales. Local decisions were set in a regional and National Forest-scale context (Cissel et al. 1998). Such an approach is being applied in National Forest planning in the Great Lakes region and elsewhere in the United States.

Moving from concept to practice in emulating natural disturbance as a guide for forest management is hampered by inadequate knowledge (Cleland et al. 2004). Disturbances occur at widely different magnitudes, frequencies, and intensities and these differences produce varied responses and outcomes. For example, at many locations within the Augusta Creek landscape, there is the possibility of low, mixed, or high fire severity, producing differences in the structure and composition of the vegetation. Natural disturbances are caused by many factors—including diseases, insects, wind, ice, extreme temperatures, fire, prolonged drought, landslides, and floods—that operate at many temporal and spatial scales. Furthermore, natural disturbances often interact with human-caused disturbances such as timber harvesting and other land uses (He et al. 2004b; Loehle 2004; Shang et al. 2004). Better understanding of the nature of these interactions is a critical need in landscape and disturbance ecology.

7.7. MANAGING ROADS

Roads are a pervasive landscape feature and are essential to our modern mobile lifestyle. There are 6.3 million km (3.9 million miles) of roads in the United States (Forman et al. 2003), the vast majority of which are public roads or private roads open to the public. Riitters and Wickham (2003) measured the proportion of land area within the conterminous United States that was located near roads of any type. Nationwide, 20% of the land area was within 127 m of a road and only 3% was more than 5176 m away. Such studies corroborate what is obvious through observation—a dense network of roads exists in most landscapes.

A host of natural resource issues such as access, remoteness, forest fragmentation, edge effects, and water quality relate to building and maintaining roads (Forman and Alexander 1998; Spellerberg 1998; Trombulak and Frissell 2000). Roads are channels for water and sediment, and are barriers to movement for some species and conduits for the dispersal of others. As road traffic and density increase, wildlife mortality increases due to vehicle–animal collisions. Roads increase the amount of edge in the landscape and decrease the amount of interior habitat. Road density is positively correlated with the level of environmental impact (Lee et al. 1997; Rieman et al. 2000). Although factors other than roads cause forest fragmentation (Heilman et al. 2002; Hessburg and Agee 2003), the relative contribution of roads to forest fragmentation is much higher in predominantly forested landscapes such as those of the Pacific Northwest or the Southern Appalachian Mountains where road densities are high (Riitters and Wickham 2003).

The USDA Forest Service manages a significant portion of the public road system in the United States—nearly 10% of its total length (Forman et al. 2003). Most forest roads are initially built for harvesting timber and are used secondarily to provide access for fire suppression, for recreational activities such as hunting and fishing, and for harvesting other forest products (e.g., mushrooms, conifer boughs for floral and wreath arrangements).

Road management by the USDA Forest Service is changing, due in part to inadequate funding to maintain the extensive current road network (as a result of declining timber harvesting in National Forests) and in part due to research on the effects of roads on terrestrial and aquatic ecosystems (e.g., Forman et al. 2003; Hann et al. 1997; Lee et al. 1997; Quigley et al. 1996; Rieman et al. 2000). In a speech to the annual conference of the Society of Environmental Journalists in September 2003, USDA Forest Service Chief Dale Bosworth stated that "for every mile of road we build, we decommission 14 miles of road. In the last 5 years, we've decommissioned 10,000 miles of road." Decommissioning roads on public lands is not easy. Once built, the public expects to use them and to continue to have access to the landscapes in which the roads exist.

Although there is a growing body of literature related to the ecological impacts of roads on both terrestrial and aquatic ecosystems and the organisms that depend on these systems, guidelines for building roads in National Forests largely reflect engineering and economic factors rather than ecological factors. Given the importance of roads to human communities and their impacts on the environment, a much more robust program of knowledge transfer is needed to balance the engineering and economic considerations with environmental concerns.

7.8. CONCLUSIONS

The value of a landscape perspective is recognized by managers and planners in the USDA Forest Service and most other resource management agencies. Scientists no longer need to convince them of its value. When viewed at the local level, no individual forest can provide all the benefits that are desired from a forest. When the local forest is viewed as part of a broader mix of forests and other land uses within the landscape, including old and young forests with varied compositions and structures, the choices are more likely to change from "either–or" to "and." Moving toward this model for resource management requires placing management decisions and actions into a more formal spatial and temporal framework (Crow and Gustafson 1997). Providing the support necessary for applying this spatial temporal framework to resource management should be a high priority among landscape ecologists. However, the metaphorical bridge that connects science with its users and represents all the mechanisms and tools for transferring information to and from users is currently far too narrow. Landscape ecologists need to deliver their knowledge in usable forms to managers, and to develop and distribute practical tools that can support the application of their emerging science.

The specific concepts and principles from landscape ecology that contribute to resource management in National Forests can be identified—managing at multiple spatial scales, relating spatial and temporal variability to the benefits derived from landscapes, and considering the ecological, economic, and social context when making local decisions (see Crow 2005 for others)—but the major contribution from landscape ecology is one of perspective. By this, I mean that this perspective supplements the view from within the forest (the common view) with a view taken from above the forest (the landscape view). When these two perspectives are combined, managers and planners have new and powerful insights available for resolving difficult problems.

The picture, however, should not be painted with too broad a brush. Scientific knowledge is but one source of information used in the decisionmaking process when managing resources. Differences also exist within and among regions in applying concepts and principles from landscape ecology to resource management. The fact remains, however, that resource managers are receptive to a landscape perspective because they perceive it to be useful for addressing pressing issues in resource management. Now it is up to scientists to deliver their science in a usable form to those wishing to apply it.

Acknowledgments

I am grateful to Ajith Perera and Lisa Buse (Ontario Ministry of Natural Resources), Janet Silbernagel (University of Wisconsin), Louise Levy (University of Minnesota), and an anonymous reviewer for their constructive comments on early drafts.

LITERATURE CITED

Aber J, Christensen N, Fernandez I, Franklin J, Hidinger L, Hunter M, MacMahon J, Mladenoff D, Pastor J, Perry D, Slangen R, van Miegroet H (2000) Applying ecological principles to management of the U.S. National Forests. Ecol Soc America, Washington, DC. Issues in Ecology, Number 6. 20 pp.

Albert DA (1995) Regional landscape ecosystems of Michigan, Minnesota and Wisconsin: a working map and classification. USDA For Serv, St. Paul, Minnesota. Gen Tech Rep NC-178.

Alig RJ, Butler BJ (2004) Projecting large-scale area changes in land use and land cover for terrestrial carbon analyses. Environ Manage 33:443–456.

Armstrong GW (1999) A stochastic characterization of the natural disturbance regime of the boreal mixedwood forest with implications for sustainable forest management. Can J For Res 29:424–433.

Bergeron Y, Harvey B (1997) Basing silviculture on natural ecosystem dynamics: an approach applied to the southern boreal mixedwood forest of Quebec. For Ecol Manage 92:235–242.

Bunting SC, Kingery JL, Hemstrom MA, Schroeder MA, Gravenmier RA, Hann WJ (2005) Altered rangeland ecosystems in the Interior Columbia Basin. In: Quigley TM (tech ed) The Interior Columbia Basin Ecosystem Management Project: Scientific Assessment. USDA For Serv, Pacific Northwest Res Stn, Portland. Gen Tech Rep PNW-GTR-553.

Chew JD, Stalling C, Moeller K (2004) Integrating knowledge for simulating vegetation change at landscape scales. West J Appl For 19:102–108.

Cissel JH, Swanson FJ, Grant GE, Olson DH, Gregory SV, Garman SL, Ashkenas LR, Hunter MG, Kertis JA, Mayo JH, McSwain MD, Swetland SG, Swindle KA, Wallin DO (1998) A landscape plan based on historical fire regimes for a managed forest ecosystem: the Augusta Creek Study. USDA For Serv, Pacific Northwest Res Stn, Portland. Gen Tech Rep PNW-GTR-422.

Cleland DT, Crow TR, Saunders SC, Dickmann DI, Maclean AL, Jordan JK, Watson RL, Sloan AM, Brosofske KD (2004) Characterizing historical and modern fire regimes in Michigan (USA): a landscape ecosystem approach. Landscape Ecol 19:311–325.

Crow TR (2002) Putting multiple use and sustained yield into a landscape context. In: Liu J, Taylor WW (eds) Integrating landscape ecology into natural resource management. Cambridge Univ Press, Cambridge, UK, pp 349–365.

Crow TR (2005) Landscape ecology and forest management. In: Wiens J, Moss M (eds) Issues and perspectives in landscape ecology. Cambridge Univ Press, Cambridge, UK, pp 201–207.

Crow TR, Gustafson EJ (1997) Ecosystem management: managing natural resources in time and space. In: Kohm K, Franklin JF (eds) Creating a forestry for the 21st century, the science of ecosystem management. Island Press, Washington, pp 215–228.

Diaz NM (2004) Northwest Forest Plan. In: Gucinski H, Miner C, Bittner B (eds) Proceedings: Views from the ridge—considerations for planning at the landscape scale. USDA For Serv, Portland. Gen Tech Rep PNW-GTR-596, pp 87–92.

Dwyer JF, Childs GM (2004) Movement of people across the landscape: a blurring of distinctions between areas, interests, and issues affecting natural resource management. Landscape Urban Planning 69:153–164.

Emanuel WR, Killough GG, Post SM, Shugart HH (1984) Modeling terrestrial ecosystems in the global carbon cycle with shifts in carbon storage capacity by land-use change. Ecology 65:970–983.

FEMAT (1993) Forest ecosystem management: an ecological, economic, and social assessment. Forest Ecosystem Management Assessment Team [USDA For Serv, USDI Fish and Wildlife Service, USDI National Park Service, USDI Bureau of Land Management, National Oceanic and Atmospheric Administration, National Marine Fisheries Services, Environmental Protection Agency]. Portland.

Finney MA (1999) Mechanistic modeling of landscape fire patterns. In: Mladenoff DJ, Baker WL (eds) Spatial modeling of forest landscape change, approaches and applications. Cambridge Univ Press, Cambridge, UK, pp 186–209.

Forman RTT (1995) Land mosaics, the ecology of landscapes and regions. Cambridge Univ Press, Cambridge, UK.

Forman RTT, Alexander LE (1998) Roads and their major ecological effects. Annu Rev Ecol Syst 29:207–231.

Forman RTT, Sperling D, Bissonette JA, Clevenger AP, Cutshall CD, Dale VH, Fahrig L, France R, Goldman CR, Heanue K, Jones JA, Swanson FJ, Turrentine T, Winter TC (2003) Road ecology, science and solutions. Island Press, Washington.

Franklin JF, Forman RTT (1987) Creating landscape patterns by forest cutting: ecological consequences and principles. Landscape Ecol 1:5–18.

GAO (2000) Report to Congressional Requestors. Forest Service planning. Better integration of broadscale assessments into forest plans is needed. United States General Accounting Office, Washington. GAO/RCED-00-56.

Gardner RH, Romme WH, Turner MG (1999) Predicting forest fire effects at landscape scales. In: Mladenoff DJ, Baker WL (eds) Spatial modeling of forest landscape change, approaches and applications. Cambridge Univ Press, Cambridge, UK, pp 163–185.

Gilmore DW, Palik BJ, Benzie JW (2004) A revised management guide for red pine. Joint Publication, USDA For Serv and Univ Minnesota, St. Paul.

Gobster PH, Rickenbach MG (2004) Private forestland parcelization and development in Wisconsin's Northwoods: perceptions of resource-oriented stakeholders. Landscape Urban Planning 69:165–182.

Gustafson EJ, Crow TR (1996) Simulating the effects of alternative forest management strategies on landscape structure. J Environ Manage 46:77–94.

Gustafson EJ, Rasmussen LV (2002) HARVEST for Windows: user's guide. USDA For Serv, North Central Res Stn, St. Paul, Minnesota. (http://www.ncrs.fs.fed.us/4153/harvest/v60/v6_0.asp)

Gustafson EJ, Sturtevant BR, Fall A (2006) A collaborative, iterative approach to transferring modeling technology to land managers. In: Perera AH, Buse LJ, Crow TR (eds) Forest landscape ecology: transferring knowledge to practice. Springer, New York.

Hammer RB, Stewart SI, Winkler RL, Radeloff VC, Voss PR (2004) Characterizing dynamic spatial and temporal residential density patterns from 1940–1990 across the North Central United States. Landscape Urban Planning 69:183–199.

Hann WJ, Jones JL, Karl MG, Hessburg PF, Keane RE, Long DG, Menakis JP, McNicoll CH, Leonard SG, Gravenmier RA, Smith BG (1997) An assessment of landscape dynamics of the Basin. In: Quigley TM, Arbelbide SJ (eds) An assessment of ecosystem components in the Interior Columbia Basin and portions of the Klamath and Great Basins, Volume 2. USDA For Serv, Pacific Northwest Res Stn, Portland. Gen Tech Rep PNW-GTR-405, pp 363–1055.

Hansen AJ, Neilson RP, Dale VH, Flather CH, Iverson LR, Currie DJ, Shafer S, Cook R, Bartlein PJ (2001) Global change in forests: responses of species, communities, and biomes. BioScience 51:765–779.

Harper SC, Falk LL, Rankin EW 1990. The northern forest lands study of New England and New York. USDA For Serv, Rutland, Vermont.

Haynes RW, Graham RT, Quigley TM (1998) A framework for ecosystem management in the Interior Columbia Basin. J For 96:4–9.

He HS, Shang BZ, Crow TR, Gustafson EJ, Shifley SR (2004a) Simulating forest fuel and fire risk dynamics across landscapes—LANDIS fuel module design. Ecol Model 180:135–151.

He HS, Shifley SR, Dijak W, Gustafson EJ (2004b) Simulating the effects of forest fire and timber harvesting on the hardwood species of central Missouri. In: Perera AJ, Buse LJ, Weber MG (eds) Emulating natural forest landscape disturbances. Concepts and applications. Columbia Univ Press, New York, pp 123–134.

Heilman GE Jr, Strittholt JR, Slosser NC, Dellasala DA (2002) Forest fragmentation of the conterminous United States: assessing forest intactness through road density and spatial characteristics. BioScience 52:411–422.

Hessburg PF, Agee JK (2003) An environmental narrative of Inland Northwest U.S. forests, 1800–1999. In: Fire and aquatic ecosystems, special feature. For Ecol Manage 178:23–59.

Hessburg PF, Smith BG, Kreiter SG, Miller CA, Salter RB, McNicoll CH, Hann WJ (1999) Historical and current forest and range landscapes in the Interior Columbia River Basin and portions of the Klamath and Great Basins, Part 1: Linking vegetation patterns and landscape vulnerability to potential insect and pathogen disturbances. USDA For Serv Portland. Gen Tech Rep PNW-GTR-458.

Hessburg PF, Reynolds KM, Salter RB, Richmond MB (2004) Using a decision support system to estimate departures of present forest landscape patterns from historical reference conditions. In: Perera AJ, Buse LJ, Weber MG (eds) Emulating natural forest landscape disturbances. Concepts and applications. Columbia Univ Press, New York, pp 158–175.

Jensen ME, Bourgeron PS (eds) (2001) A guidebook for integrated ecological assessments. Springer-Verlag, New York.

Johnson KN, Swanson F, Herring M, Greene S (eds) (1999) Bioregional assessments: science at the crossroads of management and policy. Island Press, Washington.

Keane RE, Parsons R, Hessburg PF (2002) Estimating historical range and variation of landscape patch dynamics: limitations of the simulation approach. Ecol Model 151:29–49.

Landres PB, Morgan P, Swanson FJ (1999) Overview of the use of natural variability concepts in managing ecological systems. Ecol Appl 9:1179–1188.

Lee DC, Sedell JR, Rieman BE, Thurow RF, Williams JE (1997) Broad-scale assessment of aquatic species and habitats. In: Quigley TM, Arbelbide SJ (eds) An assessment of ecosystem components in the Interior Columbia Basin and portions of the Klamath and Great Basins, Volume 2. USDA For Serv, Pacific Northwest Res Stn, Portland. Gen Tech Rep PNW-GTR-405.

Li H, Franklin JF, Swanson FJ, Spies TA (1993) Developing alternative forest cutting patterns: a simulation approach. Landscape Ecol 8:63–75.

Lindenmayer DB, Franklin JF (2002) Conserving forest biodiversity, a comprehensive multiscaled approach. Island Press, Washington.

Lint JB (tech coord) (2005) Northwest Forest Plan—the first 10 years (1994–2003): status and trend of northern spotted owl populations and habitat. USDA For Serv, Pacific Northwest Res Stn, Portland. Gen Tech Rep PNW-GTR-648.

Liu J, Taylor WW (eds) (2002) Integrating landscape ecology into natural resource management. Cambridge Univ Press, Cambridge, UK.

Loehle C (2004) Applying landscape principles to fire hazard reduction. For Ecol Manage 198:261–267.

Lytle DE, Cornett MW, Harkness MS (2006) Transferring landscape ecological knowledge in a multipartner landscape: the Border Lakes region of Minnesota and Ontario. In: Perera AH, Buse LJ, Crow TR (eds) Forest landscape ecology: transferring knowledge to practice. Springer, New York.

McGaughey RJ (1998) Techniques for visualizing the appearance of forestry operations. J For 96:9–14.

McRae DJ, Duchesne LC, Freedman B, Lynham TJ, Woodley S (2001) Comparisons between wildfire and forest harvesting and their implications to forest management. Environ Rev 9:223–260.

Mehmood SR, Zhang D (2001) Forest parcelization in the United States: a study of contributing factors. J For 99:30–34.

Moeur M, Spies TA, Hemstrom MA, Alegria J, Browning J, Cissel JH, Cohen WB, Demeo TE, Healy S, Warbington R (2005) The Northwest Forest Plan—the first ten years (1994–2003): status and trends of late-successional and old-growth forests. USDA For Serv, Pacific Northwest Res Stn, Portland. Gen Tech Rep PNW-GTR-646.

Muhar A (2001) Three-dimensional modeling and visualization of vegetation for landscape simulation. Landscape Urban Planning 54:5–17.

Neilson RP (1995) A model for predicting continental-scale vegetation distribution and water balance. Ecol Appl 5:362-385.

Pennock DJ, van Kessel C (1997) Effect of agriculture and of clear-cut forest harvest on landscape-scale soil organic carbon storage in Saskatchewan. Can J Soil Sci 77:211–218.

Perera AH, Buse LJ, Weber MG (eds) (2004) Emulating natural forest landscape disturbances. Concepts and applications. Columbia Univ Press, New York.

Quigley TM, Arbelbide SJ (eds) (1997) An assessment of ecosystem components in the Interior Columbia Basin and portions of the Klamath and Great Basins. Volume 2. USDA For Serv, Pacific Northwest Res Stn, Portland. Gen Tech Rep PNW-GTR-405.

Quigley TM, Haynes RW, Graham RT (eds) (1996) Integrated scientific assessment for ecosystem management in the Interior Columbia Basin and portions of the Klamath and Great Basins. USDA For Serv, Pacific Northwest Res Stn, Portland. Gen Tech Rep PNW-GTR-382.

Rieman BE, Lee DC, Thurow RF, Hessburg PF, Sedell JR (2000) Toward an integrated classification of ecosystems: defining opportunities for managing fish and forest health. Environ Manage 25:425–444.

Riitters KH, Wickham JD (2003) How far to the nearest road? Front Ecol Environ 1:125–129.

Riitters K, Wickham J, O'Neill R, Jones B, Smith ER (2000) Global-scale patterns of forest fragmentation. Conserv Ecol 4(2):3. (http://www.consecol.org/vol4/iss2/art3)

Riitters KH, Wickham JD, O'Neill RV, Jones KB, Smith ER, Coulston JW, Wade TG, Smith JH (2002) Fragmentation of continental United States forests. Ecosystems 5:815–822.

Schulte LA, Mladenoff DJ (2001) The original US public land survey records: their use and limitations in reconstructing presettlement vegetation. J For 99:5–10.

Schulte LA, Crow TR, Vissage J, Cleland DT (2003) Seventy years of forest change in the northern Great Lakes Region, USA. In: Buse L, Perera A (comps), Proceedings, Regional Forestry Forum. Ont Min Nat Resour, Ont For Res Inst, Sault Sainte Marie, Ont For Res Inf Pap Number 155, pp 99–101.

Shang BZ, He HS, Crow TR, Shifley SR (2004) Fuel load reductions and fire risk in central hardwood forests of the United States: a spatial simulation study. Ecol Model 180:89–102.

Spellerberg IF (1998) Ecological effects of roads and traffic: a literature review. Global Ecol Biogeogr 7:317–333.

Spies TA (1998) Coastal landscape analysis and modeling study (CLAMS) program report. Oregon State Univ, Corvallis.

Spies TA, Ripple WJ, Bradshaw GA (1994) Dynamics and pattern of a managed coniferous forest landscape in Oregon. Ecol Appl 4:555–568.

Stewart WP, Liebert D, Larkin KW (2004) Community identities as visions for landscape change. Landscape Urban Planning 69:315–334.

Swanson FJ, Franklin JF, Sedell JR (1997) Landscape patterns, disturbance, and management in the Pacific Northwest, USA. In: Zonneveld IS, Forman RTT (eds) Changing landscapes: an ecological perspective. Springer-Verlag, New York, pp 191–213.

Swetnam TW, Betancourt JL (1990) Fire–southern oscillation relations in the southwestern United States. Science 249:1017–1020.

Trombulak SC, Frissell CA (2000) Review of ecological effects of roads on terrestrial and aquatic communities. Conserv Biol 14:18–30.

USDA Forest Service (2001) 2000 RPA assessment of forest and range lands. USDA For Serv, Washington. FS-687.

Wade TG, Riitters KH, Wickham JD, Jones KB (2003) Distribution and causes of global forest fragmentation. Conserv Ecol 7(2):7. (http://www.ecologyandsociety.org/vol7/iss2/art7/)

Wallin DO, Swanson FJ, Marks B, Cissel JH, Kertis J (1996) Comparison of managed and pre-settlement landscape dynamics in forests of the Pacific Northwest, USA. For Ecol Manage 85:291–309.

Wang X, Song B, Chen J, Zheng D, Crow TR (2006) Visualizing forest landscapes using public data sources. Landscape Urban Planning 75: 111–124.

Wear DN, Greis JG (2002) Southern Forest Resource Assessment, summary of findings. J For 100:6–14.

Wear DN, Greis JG (2003) The Southern Forest Resource Assessment: final report. USDA For Serv, Southern Res Stn, Asheville, North Carolina. (www.srs.fs.fed.us/sustain/report/index.htm)

Wiens JA, Moss MR (2005) Issues and perspectives in landscape ecology. Cambridge Univ Press, Cambridge, UK.

Wimberly MC, Spies TA, Nonaka E (2004) Using criteria based on the natural fire regime to evaluate forest management in the Oregon Coast Range of the United States. In: Perera AH, Buse LJ, Weber MG (eds) Emulating natural forest landscape disturbances. Concepts and applications. Columbia Univ Press, New York, pp 146–157.

Zasada JC, Palik BJ, Crow TR, Gilmore DW (2004) Emulating natural forest disturbance, applications for silvicultural systems in the northern Great Lakes region of the United States. In: Perera AH, Buse LJ, Weber MG (eds) Emulating natural forest landscape disturbances. Concepts and applications. Columbia Univ Press, New York, pp 230–242.

Zuuring HR, Wood WL, Jones JG (1995) Overview of MAGIS: a multi-resource analysis and geographic information system. USDA For Serv, Intermountain Res Stn, Ogden, Utah. Res Note INT-RN-427.

8

Fundamentals of Knowledge Transfer and Extension

A. Scott Reed and Viviane Simon-Brown

8.1. Introduction
8.2. Definitions of Key Terms
8.3. A Common Philosophical Foundation
    8.3.1. Engagement
    8.3.2. Relationships
    8.3.3. Scholarship
8.4. An Empowering Environment
    8.4.1. Knowledge Transfer Is a Choice
    8.4.2. It Requires Both Stability of Funds and Nimbleness
    8.4.3. There Are Structural Implications
    8.4.4. Accountability and Evaluation Are Included
    8.4.5. Skilled Professionals Are Hired and Supported
8.5. Effective Program- or Activity-Design Principles
    8.5.1. Learner-Centric Education
    8.5.2. Credible Research-Based Information
    8.5.3. Evaluation
8.6. Summary and Ongoing Challenges
Literature Cited

8.1. INTRODUCTION

Increasingly, forestry researchers must do more than solve abstract problems; they must also make those solutions available to those who can use them. Research results and other sources of innovation fall far short of their potential to change management practice when these resources do not become part of the working knowledge of those

---

A. SCOTT REED and VIVIANE SIMON-BROWN • Oregon State University, College of Forestry Extension Service, Richardson Hall 109, Corvallis, OR 97331, USA.

who will make the changes. And, once these resources are applied, the resulting adaptive management raises new questions that drive further research and knowledge development. To ensure that this iterative process occurs effectively, a process for accomplishing as many as possible of the following goals is necessary:

- Application of research results: Through effective transfer of new knowledge and methodologies, research results become part of operational practice.
- Validation of new knowledge: Practitioners ensure that research results are realistic based on their years of experience.
- Operational testing: Practitioners test new practices and new information at an operational level to confirm whether the results are operationally significant.
- Feedback that can help prioritize additional research: After applying new knowledge and practices, practitioners provide feedback to researchers to guide their future research and help them refine new research hypotheses.
- Ongoing dialogue: Knowledge generators, knowledge transfer professionals, and those who apply the knowledge interact through systematic, designed or informal interactions, thereby generating new innovations that would not otherwise have occurred.

The continuum from the development to the application of knowledge involves three major players (DeYoe and Hollstedt 2003): researchers, knowledge transfer (or "extension") professionals, and practitioners. A fourth major category of player, the citizens and communities in which forestry activities occur, cannot be neglected. As such, we mention the role of the public periodically throughout this chapter.

Researchers generate and develop knowledge. They define problems, identify desired outcomes, plan their approach, conduct basic and applied research, and explore the development possibilities. Their work can thus be described as thinking, seeking answers, questioning and formulating hypotheses, testing hypotheses, assessing and interpreting the results of their studies, and publishing the results. They may or may not develop the technology permitted by these results or carry out pilot testing and ground-truthing. They generally communicate mostly within the scientific community.

On the other end of the continuum, practitioners operationalize the researcher's work. They are actively engaged in communication within their organization and with key stakeholders, in operational testing and implementation of new approaches, and in conducting trials of adaptive management. Based on the results of this work, they may evaluate efficacy, supply innovations that modify an original concept, and provide feedback to those who proposed that concept. Finally, they adopt new knowledge and technologies, and either develop new policies and practices or revise old ones. In short, they act and implement while responding to the issues and deadlines that govern their work, and weigh contingencies and risks in so doing.

Knowledge transfer professionals complete the continuum by bridging the gap between those who generate the knowledge and those who apply it. To do so, they engage in audience education and training by linking traditional scientific and operational knowledge with new discoveries, by collecting and synthesizing information, and by demonstrating techniques or conducting operational testing in close cooperation with practitioners. These professionals employ a variety of strategies and technologies to accomplish these goals, and help the audience to identify their needs and any new research and development capable of meeting those needs. And they communicate, facilitate, mediate, synthesize, simplify, and act as liaisons between researchers and practitioners. In addition, they provide outreach and troubleshooting services once the researchers and practitioners have begun to interact.

Institutions that engage in the transfer of knowledge to the forestry community have highly variable organizational structures, but share the common goals of helping both researchers and practitioners to solve problems, manage their resources, and better engage—all of which contribute to the long-term sustainability of forests and provision of their many benefits.

Peter Bloome, professor emeritus at Oregon State University, has proposed that successful knowledge transfer depends on three principles:

- responsiveness to locally identified issues, which helps to ensure an audience's receptivity toward educational activities that address their expressed needs;
- well-informed citizens capable of contributing to sound community decisions; and
- the achievement of broad social goals through the development of relationships and encouragement of communication among those who share a common vision.

Although these principles are directed at a citizen audience, they can be generalized for other audiences. Based on our own experience, three general categories of key factors are required for successful transfer: a common philosophical foundation, an empowering institutional environment, and effective design principles. Although each category is critically important, we have paid special attention in this chapter to the third category, for which those engaging in knowledge transfer individually have the most direct influence. As transfer professionals, we are familiar with a formalized transfer program so that is our focus here. However, we also summarize the associated principles to help landscape ecologists recognize and apply the principles and approaches outside a formal program.

The broad goal of this chapter is to systematically describe the elements of effective knowledge transfer that match important information with receptive learners who can use that information. Effective knowledge transfer is predicated on (1) engaging these learners in ways that make the educational content clearly relevant to their circumstances, (2) building alliances among individuals and organizations with shared goals, and (3) working to adapt both past experience and new knowledge to improve operational practice. The specific goals of this chapter are thus to suggest a common set of terms and definitions that describe the key elements of knowledge transfer; describe the key elements of successful knowledge transfer activities; illustrate the essential skills for developing, implementing, and evaluating transfer activities; and describe some challenges of adapting knowledge transfer to changing circumstances.

8.2. DEFINITIONS OF KEY TERMS

Definitions of many of the key terms in the knowledge transfer process vary. To ensure consistency, we have chosen to define the terms that we will be using in this chapter as follows: *Knowledge transfer* includes any effort to deliver knowledge to someone who wishes to or needs to receive it, including the publication of research for use by the scientific and practitioner communities. Knowledge transfer is a precursor to *technology transfer*, which involves the transfer of the results of basic and applied research to the design, development, and commercialization of new or improved products, tools, services, or processes. In both cases, the recipients of this transfer are called the *audience*; other names include *clients*, *students*, or *customers*, but in this chapter, we will use only the more inclusive term. Others with an interest in the outcomes of the knowledge transfer, whether or not they are themselves part of the audience, are called *stakeholders*.

The overall knowledge transfer process is sometimes referred to as *extension*, particularly when the organization responsible for this task is based at an American Land Grant University, whose mission goes beyond educating the university's students. When the knowledge or technology transfer is part of an organized approach, we have used the term *program*. A program is more than a specific activity; rather, it includes an assortment of activities, associated materials, and learning activities related to a specific topic or educational goal.

8.3. A COMMON PHILOSOPHICAL FOUNDATION

The philosophy of knowledge transfer rests on a common foundation, with three main dimensions, defined here and expanded below:

- Engagement: Audiences must be engaged in (involved in) customizing their learning experience.
- Relationships: Transfer occurs via human interactions among individuals, their communities, and their respective organizations.
- Scholarship: High-quality knowledge transfer activities must meet a high standard of excellence, and thus must incorporate some measure of peer review and validation and the possibility of replication by others.

8.3.1. Engagement

As defined by the Kellogg Commission (1998), *engagement* involves the audience as an active participant in the learning experience. This involvement takes several forms, including actively defining the issues and problems and interacting with the

Fundamentals of Knowledge Transfer and Extension

educators (transfer specialists) to enhance knowledge transfer and establish a co-learning environment in which both the educator and the audience benefit. Engagement enriches the learning experience by enhancing opportunities for researchers to develop active relationships with their audience.

Several factors can enhance the level of engagement in support of successful transfer: responsiveness, respect for audience, neutrality, accessibility, integration, coordination, and resource partnerships. These are described in more detail below, and those that may be more relevant to programs are identified.

8.3.1.1. Responsiveness

Those conducting transfer, be they researchers or transfer specialists, can promote engagement by being responsive to their audience and thereby ensuring that transfer activities are relevant. A responsive transfer program or activity asks the right questions, offers effective and timely services, and engages with the audience in the following ways:

- It asks questions that help define the real problems and the real constraints on solving those problems.
- It offers services in a useful format and at the appropriate time.
- It ensures that communications are clear.
- It requests input from stakeholders.
- It invests in open discussions to best understand the dimensions of the problem or issue.
- It understands that by reaching out, valuable information for program development will be obtained.

8.3.1.2. Respect for Audience

The fundamental purpose of engagement is not to provide the researcher's or transfer specialist's superior expertise to a less-competent audience, but rather to encourage joint definition of problems, solutions, and criteria for success. In essence, this means respecting the audience. Respect for an audience is shown by:

- Genuinely expressing appreciation and respect for the skills and capacities of partners in collaborative projects.
- Involving those people who will be affected by our decisions and any program that results from these decisions.
- Showing that we have as much to learn as we have to offer.

8.3.1.3. Neutrality

Of necessity, some of our transfer activities will involve contentious issues, for which multiple "right" answers exist. Remaining objective and offering alternative solutions often best meets audience needs because it allows the transfer specialist to act as a neutral arbitrator between the various stakeholders. Neutrality includes:

- Maintaining our role as an unbiased facilitator of learning and consideration of alternatives.
- Managing an environment in which participants feel comfortable exchanging ideas.

Although the principles outlined below refer primarily to more formal transfer programs, the underlying principles also are relevant to those contemplating less formal transfer activities.

8.3.1.4. Accessibility

The institutions created by transfer specialists are often confusing to outsiders. To resolve this confusion, we need to find ways to help inexperienced audiences understand and negotiate complex structures so that what we have to offer is readily available. To gauge accessibility, we should consider:

- Wide and appropriate publicity of activities and resources.
- Accommodation of those with special access needs.
- Offering a variety of formats to ensure participation.

8.3.1.5. Integration

We must find a way to integrate our service with our audience (and the public, if they are not formally part of that audience), as our responsibility is to develop and share our intellectual capital. This is best accomplished by fitting transfer efforts into existing systems through integration. To succeed, the institutional climate should foster outreach, service, and engagement. A commitment to interdisciplinary and interorganizational work is indispensable within an integrated approach. Integration considers:

- Incentives that are useful in encouraging researchers, transfer specialists and audiences to effectively engage.
- Respected and senior staff leaders not only participate but serve as advocates for knowledge transfer.
- Enlistment of other organizations or individuals who can contribute to the process.

8.3.1.6. Coordination

When integration is achieved, coordination becomes an issue: someone must take responsibility for ensuring that all parties are involved, cooperative, and aware of the efforts of the other parties. The task of coordinating activities—whether through serving a management role, creating advisory councils, or providing thematic structures such as multidisciplinary institutes or centers—clearly requires considerable attention. Appropriate coordination means:

- Parties are dealing with each other productively.
- The goal of engagement is understood.
- The need for any party to develop knowledge transfer skills is recognized and addressed.

8.3.1.7. Resource Partnerships

The final test asks whether the resources committed to the task are sufficient. Engagement is not free; the time and effort of participants and the development and implementation of activities all have costs. The most successful engagement efforts are associated with strong and healthy partnerships that ensure the availability of appropriate resources. This adequacy of resources is evaluated by:

- Availability of funding.
- Potential for corporate sponsorship and investment.
- Potential for alliances and strategic partnerships to be formed between government and industry.
- Determining whether new fee structures can be developed for delivery of services.

8.3.2. Relationships

Building relationships is an inevitable outgrowth of engagement. There are several reasons why recognizing and consciously promoting relationships makes sense. First, the involvement of multiple individuals and organizations increases visibility of the issue. This may help to draw additional stakeholders into the learning environment. Second, partnerships, by their nature, lead to commitments that can promote organizational action that, in their absence, might only become rhetoric. Third, the base of skills and resources available to address a problem often increases due to the skills and energy of the additional members.

These relationships are a key factor in expanding a sense of ownership and commitment to working together on common issues. The nature of the relationships may be characterized in several ways:

- Communities: Increased interest in a challenge such as ecosystem management emphasizes the need to engage multiple stakeholders and include the communities within various geographic regions (such as watersheds) in the development and implementation of sustainable management practices for the benefit of the ecosystem and the stakeholders.

- Other institutions: The effectiveness of problem-solving is enhanced by establishing partnerships with other organizations that can provide the benefits of increased scope and scale. The private sector is particularly able to adapt and expand technology transfer to produce marketable products with economic benefits.
- Teams: Few contemporary problems are simple enough to be addressed by a single specialist. Therefore, knowledge transfer activities are enhanced by encouraging the formation of teams that combine the strengths of several disciplines to bolster the content, design, and delivery of one or more activities.
- Stakeholders: Many organizations and individuals share an interest in the success of transfer activities. Involvement of these stakeholders reinforces their mutual concern and their investment of energy, time, and resources.

Distinguishing different types of linkages contributes to understanding the nature of various relationships and the expectations of each participant. These linkages can be defined in terms of three levels of increasing complexity and engagement: cooperation, collaboration, and partnership (Table 8.1), ranging from low-risk or no-risk relationships to fully interdependent linkages (Hogue and Miller 2000). Each has a different purpose, structure, and process for accomplishing its goals (Bergstrom et al. 1996).

8.3.2.1. Cooperation

Cooperation is the least-demanding form of linkage; in its simplest form, it may involve nothing more complex than an informal agreement to avoid interference with each cooperator's goals and activities. However, a true cooperative relationship typically involves sharing of an activity, campaign, or event between organizations as a result of an invitation from one organization to another. The request is seen as consistent with that organization's mission, values, and goals.

Management of the cooperative structure is centralized by means of an informal or semiformal coordinating body. Communication is somewhat centralized and

Table 8.1. Summary of the three main levels of linkage among knowledge transfer participants showing increasing complexity, from cooperation to partnership

Main levels of linkage		
Cooperation	Collaboration	Partnership
• Shared activity at the request of one organization	• Overlapping missions	• A new entity in which former organizational identities are deemphasized
• Relatively short-term; informally defined roles	• Shared resources; active teamwork	• High levels of trust and integration of activities
• Defined, short-term, informal or semiformal organizational arrangement	• Formal relationship defined at high levels	• Decisions by consensus

formal, as group members generally have little or no history of working together. To accomplish the cooperative's intended goals, group members either provide money from their own organizations, or undertake a communitywide fundraising effort.

The process used in this type of linkage is relatively simple. At the beginning of the activity planning, group members select one or two leaders. These leaders allow group members to make numerous interconnected decisions for a range of tasks. The leaders also strive to reduce interpersonal conflicts within the group. Once an activity or program has been completed, the cooperative disbands.

8.3.2.2. Collaboration

In a collaborative relationship, the participating organizations share resources rather than just an activity, use the existing resource base to create new resources, and develop new bases of support that benefit the participating organizations as well as the collaboration itself (Bergstrom et al. 1996). The missions of the organizations generally overlap to some extent, and the partners accomplish the mutual portion of their missions through joint planning. Ownership and credit for the activities is shared equally. This type of linkage is less common than simple cooperation because of the perceived or real relinquishment of each organization's unique identity.

A collaboration is best made operational through prenegotiation and written understandings, and is thus more formal than a cooperative. The structure, including roles, responsibilities, and decisionmaking criteria, is formally defined. Because the collaboration uses significant resources from the parent organizations, the people at the hub of the collaboration are generally higher-level decisionmakers. Most collaborations create a joint budget from newly developed and existing resources. Members communicate frequently and clearly through formal hierarchical channels. Decisionmaking can take place at multiple, previously agreed-upon levels. The collaboration leaders often act independently of their primary organization.

8.3.2.3. Partnership

True partnerships are relatively rare. In this relationship, the participating groups create a new system, with the identities of the individual organizations being subsumed into a new entity. True partnerships are trust-based; partners do not disadvantage each other. Even though a new organization is formed, the history and culture of the parent organizations is still valued and their strengths are encouraged to flourish. The purpose of a true partnership is to work toward a shared vision and mission with tangible results and identifiable impacts. The participating groups develop sophisticated and interdependent systems of ongoing support, including funding, staffing, and operations. Consensus is the preferred decisionmaking method. The relationship is formalized by means of memoranda of understanding and often by statutes that ensure a nonprofit operating structure. The newly formed organization defines work plans and assignments, prescribes roles and responsibilities, and specifies reporting and evaluation criteria.

The nature of a true partnership ensures high levels of trust and productivity. Innovative leadership is the norm. The internal processes of this type of linkage are highly developed and productive, with ideas and decisions being shared equally.

The challenge for those undertaking transfer activities is to consciously choose the level of engagement (cooperation, collaboration, partnership) that best suits a given situation and set of transfer objectives.

8.3.3. Scholarship

Effective knowledge transfer is best done by skilled professionals who have specialized credentials and experience in the design, delivery, and evaluation of knowledge transfer activities. Few academic institutions provide students with coursework designed to produce transfer professionals. Many of our colleagues working in this field have developed their abilities through active involvement in knowledge transfer activities and by developing working relationships with more experienced mentors. Regardless of how their skills are obtained, knowledge transfer professionals learn to plan for desired outcomes, accommodate the attributes of the audience, and implement known instructional design principles. Others contemplating transfer activities can learn from their experiences.

The simplest goal of scholarship is to produce activities and products of high "quality," but quality must be defined. The traditional culture of the research community is that results are exposed to the scrutiny of peers. The resulting peer-reviewed information is generally regarded as having met a higher standard than would be the case if the individual researcher simply provided their own interpretations and conclusions. An even higher standard involves peer-refereeing, in which the referees have the option of declaring certain results unpublishable or certain conclusions indefensible because of deficiencies in the methodology or rigor of the study.

Most current thinking about scholarly work related to knowledge transfer activities springs from the work of Boyer (1990), who argued for a significant expansion of simple research (the "scholarship of discovery"). At Oregon State University, for example, scholarship is required from faculty members who wish to establish tenure and be promoted to higher faculty ranks (OSU 2002). The same expectation applies to teachers, transfer professionals, and researchers who are asked to (1) create something that is new or innovative, (2) accomplish validation of its quality through peers, and (3) appropriately share and archive the contribution to ensure access to it by other scholars.

Among the attributes of university-based knowledge transfer personnel is the expectation that a scholarly approach will strengthen the design, delivery, and evaluation of educational activities. To support these goals, Oregon State University has developed a simple, three-part definition of scholarship that includes the creation of something new or innovative, validation of the results by peers, and appropriate documentation and archiving of the results. Opportunities exist to better define scholarly activities that would advance the concept of engagement and for others contemplating knowledge transfer to learn from these.

8.4. AN EMPOWERING ENVIRONMENT

Whether the organization involved in knowledge transfer is public or private, certain features help to support a durable commitment to the knowledge transfer function: making it a policy choice for organizations and a conscious choice for individuals, ensuring dedicated funding, allowing organizational flexibility, being accountable for and evaluating outcomes, and hiring skilled professionals or acquiring relevant skills. Although the discussion in the remainder of this section refers mainly to formal programs, once again the principles apply to those engaging in knowledge transfer activities outside a formal program.

8.4.1. Knowledge Transfer Is a Choice

Knowledge transfer is one approach to stimulating the behavioral changes required to achieve some goal, such as applying research knowledge or improving the public welfare. Figure 8.1 illustrates how social goals combine with audience characteristics to inform a policy or activity such as education, technical assistance, or regulation. Programs undertake knowledge transfer to achieve desired results. Evaluation of the returns on such investments may be considered in classical terms of efficiency (i.e., return on investment) and equity (i.e., the extent to which benefits are shared among stakeholders).

8.4.2. It Requires Both Stability of Funds and Nimbleness

A durable institutional commitment provides stability of funding over a relatively long period of time, allowing participants to maintain ongoing relationships with their audience. Such a commitment allows for ongoing professional development and the refinement of a knowledge base through the associated development of

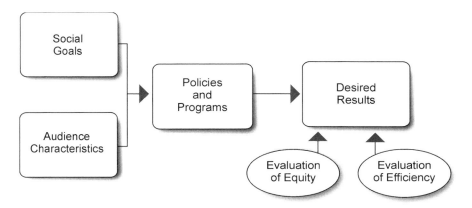

Figure 8.1. The interaction of social goals and audience characteristics to shape policy and produce the desired results. Transfer occurs between the boxes.

knowledge. However, many educational issues that require additional knowledge are more critical and shorter-term in nature, and these can often be difficult to fund from the more durable, long-term kind of funding. Thus, the organization must be able to attract new fixed-term resources that permit immediate attention to the problem, or must be sufficiently nimble to reallocate funds as well as effort to deal with these problems. Leveraging long-term resources allows a more nimble response to short-term, high priority needs. This combination of a commitment to sustained provision of resources for fundamental activities while permitting nimbleness in response to emerging issues ensures both support for long-term transfer projects and responsiveness to unforeseen transfer needs.

8.4.3. There Are Structural Implications

Knowledge transfer organizations must learn to apply the benefits of long-term resources and short-term flexibility to create an organizational structure capable of making progress on critical long-term problems that face their audiences, yet without preventing responses to unexpected problems. The ease of organizational change is influenced by the rigidity of the organization itself. Characteristics that support a more adaptable organization include active use of strategic planning to implement new initiatives, routine deployment of special project teams when necessary, and allowing for creative adaptation and flexibility in setting work priorities.

8.4.4. Accountability and Evaluation Are Included

Organizations and individuals are increasingly asked to account for their use of the resources and support provided by various stakeholders. Ongoing use of performance-based metrics helps to build confidence in and support for the knowledge transfer organization. Measures of accomplishment can be characterized as inputs, outputs, and outcomes. *Inputs* are measures of the amount of resources, including worker time and funds, that have been invested. *Outputs* describe the knowledge transfer activities, the people reached, and their initial reactions to the activities. *Outcomes* describe the social, environmental, or economic impacts of these outputs, and are regarded as the most powerful evidence of success. However, they also are usually the most difficult metric to produce. Evaluation is necessary both for accountability purposes and for continuous improvement of the effectiveness of transfer efforts.

8.4.5. Skilled Professionals Are Hired and Supported

Research and knowledge transfer require different skills. Individual researchers are expected to be expert in their area of research, but are not expected to have advanced knowledge transfer skills in addition to the required research competencies. These skills *can* be learned, and many gifted knowledge transfer professionals develop expertise in research, knowledge transfer, and application of the knowledge.

However the necessary skills are obtained, successful and effective knowledge transfer requires the organization to hire (or train) skilled professionals, and to provide them with the support they require to apply their skills. Suitable individuals possess the following characteristics, summarized here and outlined in more detail below:

- awareness of the roles within the knowledge transfer system
- appropriate knowledge of discipline
- knowledge of instructional-design tools
- exceptional interpersonal communication skills, and
- personal character traits that enhance knowledge transfer

8.4.5.1. Awareness of the Roles within the Knowledge Transfer System

Researchers generate knowledge that practitioners will want to use. They may transfer their own knowledge or work with knowledge transfer professionals whose role is to facilitate the exchange of knowledge between those who create it and those who will use it.

An example illustrates the interactions among these various roles: A group of researchers studied the effects of six alternative management scenarios on forest succession. They developed the study methods, created the simulation approach, obtained the necessary input data, analyzed the model outputs, and discussed the results—all classic researcher roles. When they published the results in the research journal *Landscape Ecology* (Gustafson et al. 2004), they began the transfer of knowledge to the scientific community, and ultimately to the people in the field who could benefit from their work. The researchers can choose to continue the knowledge transfer process themselves, can seek assistance from skilled professionals who will share the responsibility, or can return to the research arena and hope that the published information will eventually be picked up and used by practitioners. The knowledge transfer specialist can take on the role of facilitating the first option, of performing the second option, or of persuading the researchers that the third option is not the most effective approach.

8.4.5.2. Appropriate Knowledge of a Discipline

Knowledge transfer requires an understanding of the discipline underlying what is being transferred. However, the ability to integrate and synthesize information from multiple sources and disciplines may be more important than deep expertise in a single subject. Subject-matter expertise must also be combined with expertise in the delivery of transfer activities.

Knowledge transfer professionals benefit from grounding in several fields and from understanding the concepts and methodologies of systems-based thinking to address complex situations. Rather than reducing a complex situation into its simplest parts, a systems approach to thinking about problems recognizes and attempts to deal with the complexity of the whole. This ability to conceptualize the big picture helps audience members to develop new ways of thinking about problems (Patterson 1991). It also presents opportunities for spontaneous innovations that can advance the field. These breakthroughs can occur when audience members from different disciplines engage in collaborative thinking and conversation about the desired outcomes, or when the systems thinking triggers a "eureka" moment in a participant (DeYoe and Hollstedt 2003).

8.4.5.3. Knowledge of Instructional-Design Tools

In the context of most forestry problems, knowledge transfer involves adult education. To succeed in this type of education, transfer specialists must understand who the learners are, what they need to learn, and how they learn (Norland 2003). Understanding current theories of adult learning helps in planning, implementing, and evaluating activities. Knowledge transfer thus depends on delivering high-quality, timely educational experiences tailored to the needs and abilities of adult learners (Reed 1999). Patterson (1991) argues that professionals should have thorough understanding of their own learning processes so they can best facilitate the learning of others. The term "autonomous learner" encompasses this concept by defining a professional who possesses both subject-matter expertise and the ability to manage information and new experiences so as to solve problems and make decisions. The transfer specialist requires keen diagnostic skills to ascertain the audience's learning needs, an ability to respond quickly to changing situations, and a predisposition toward encouraging knowledge "exchange."

8.4.5.4. Exceptional Interpersonal Communications Skills

Effective communication is the core characteristic of a knowledge transfer professional. Talented communicators understand people and have learned to get the message across in ways that enable their audience to learn. Transfer specialists have knowledge of a variety of interpersonal and public communication techniques, and can adapt them to suit each learner's needs. They are attuned to people and their environment, and can communicate effectively both orally and in writing with individuals, small groups, and large groups. Active listening is one of the most useful communication tools in their repertoire, as this approach acknowledges the dual roles of learner and teacher (CRC 1998).

8.4.5.5. Personal Character Traits that Enhance Knowledge Transfer

Certain character traits distinguish the most effective knowledge transfer professionals. As experts, they are dependable, fair, honest, and trustworthy, and demonstrate strong teamwork and people skills. They respond promptly to audience needs. As a result, they are highly credible and are respected for their knowledge. They also appreciate the difference between knowledge and wisdom: knowledge is something learned, but wisdom is knowledge that has been tempered by the test of time and real-world application (Fletcher 1999). They stay current in their fields, and ensure that their activities evolve to meet changing audience needs. They are enthusiastically committed to their subject and to the knowledge transfer process, maintain positive attitudes, and are accepted by their audience as trusted partners, and perhaps even friends (Cooper and Graham 2001). They exhibit genuine customer-service ethics, and deeply wish to improve the public good.

Skilled knowledge transfer professionals engender trust, both through communication skills and through commitment to developing and maintaining ongoing working relationships with an audience. They respect an audience's skills, experiences, and knowledge. The long-term allegiances that result are one of the distinguishing characteristics of successful knowledge transfer.

Such knowledge transfer professionals can make an incredible contribution to science, target audiences, and the public good. The following example illustrates the breadth, depth, diversity, and range of one university-based knowledge transfer specialist who has worked as an extension forester for 19 years. In terms of knowledge transfer, this person has organized 375 events that attracted more than 20 000 participants, and has provided informal, one-on-one assistance to more than 12 000 individuals. In terms of research and scholarship, he initiated three research studies, and authored or coauthored 33 extension publications plus 13 scientific papers. In terms of grants and contracts, the person was the principal investigator or coinvestigator in 17 competitive grants, and enabled the donation of a 120-acre forest property for a total of USD$340 500. Additional tangible impacts of this activity include the fact that two decades ago, most landowners viewed red alder trees (*Alnus rubra*) as unmarketable. As a result of this extension forester's red alder research, this tree is now managed as a commercially valuable species. This knowledge transfer specialist organized a Christmas tree marketing association; more than 40 growers pooled their resources, resulting in $600 000 in farm gate sales. He played a key role in the development and delivery of the original Master Woodland Manager program, which has expanded to 20 states and 10 countries, underscoring the importance of persistence and long-term commitment for successful transfer.

8.5. EFFECTIVE PROGRAM- OR ACTIVITY-DESIGN PRINCIPLES

What constitutes "best practices" in knowledge transfer? The three fundamental attributes are that the program or activity is learner-centric, relies on credible research-based information, and is followed by a rigorous evaluation to ensure continuous improvement.

8.5.1. Learner-Centric Education

Effective knowledge transfer does not focus on what the educator wants to teach; it focuses on what the learner needs in order to develop appropriate knowledge, skills, attitudes, and behaviors. The best activities or programs happen when the audience and the knowledge transfer professional freely exchange information, experiences, and problem-solving insights. As illustrated in Figure 8.2 and expanded below, the essential steps in developing learner-centric education are to identify audience needs, create a positive learning environment, incorporate a range of teaching modalities to accommodate different learning styles, adapt to the independent, self-directed nature of adult learners, adopt a minimalist philosophy, and document personal and group achievements.

8.5.1.1. Identify Audience Needs

Excellent knowledge transfer begins with identifying the issues (DeYoe and Hollstedt 2003). The philosophical foundations of engagement and relationships consistently present opportunities for input from audiences. Knowledge transfer specialists are committed to learning about the audience's perceptions of issues and trends, their current needs for educational activities, and what they might wish to see offered in the near future (Reichenbach and Simon-Brown 2002).

Approaches for identifying audience needs can be proposed by stakeholders, policymakers, researchers, and knowledge transfer professionals. Informal ways of garnering information include individual telephone calls, taking advantage of unplanned encounters, or responding to unsolicited e-mails from an audience member. Focus groups, interviews, and surveys comprise more formal inquiry methods.

Stakeholders may be engaged on a one-time basis, or may be part of established, semipermanent advisory or working committees (Johnson 2003). Addressing

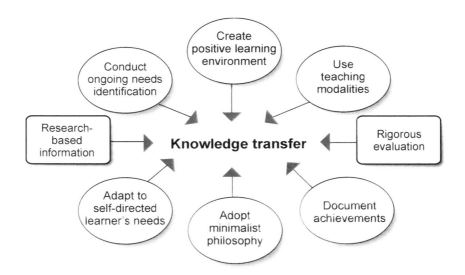

Figure 8.2. The essential components of knowledge transfer process.

the needs of both categories of stakeholder increases the likelihood of the activity meeting the audience's needs. With existing audiences, stakeholders should be involved in the entire design process from needs assessment through implementation and evaluation (Johnson 2003).

This approach works well with known audiences, but it is far more difficult to connect with new audiences to address their emerging issues. Identifying and contacting interest groups, corporations, professional and nonprofit organizations, resource users, community and political leaders, and other education professionals are effective means of expanding beyond the traditional audience base.

This ongoing effort to identify the needs of the audience can reveal additional issues requiring transfer. In that situation, prioritizing new requests for activities becomes necessary, since human and financial resources are limited. This step involves careful scrutiny—analyzing strengths, weaknesses, opportunities, and threats (SWOT)—and an internal assessment that revisits the organization's philosophical foundations and ensures that its strategic goals and activities are consistent; this scrutiny can help direct the group's energies (Simon-Brown 1999). As well, distinguishing between what audiences say they want and what they actually need is an important skill, and this scrutiny can help to set transfer priorities.

8.5.1.2. Create a Positive Learning Environment

Creating a safe and motivating intellectual environment for a thoughtful exploration of knowledge is the key to personalizing learning. To do so, we must consciously think about the needs of the learners. Physical well-being is similarly important. For example, if the audience is attending after-work classes, then comfortable seating, access to refreshments, and appropriate class durations (neither too long nor too short) all become important components of the knowledge transfer. For American contexts, meeting the spirit as well as the letter of the law of the Americans with Disabilities Act (ADA) is also crucial (Simon-Brown 1999); even where no such legislation exists, we should honor the spirit of accommodating the range of needs of all audience members.

Adult learners arrive with broad and diverse sets of prior knowledge, values, beliefs, and life experiences—both positive and negative—that influence their ability to learn (Norland 2003). For many adults, a teacher lecturing in front of a classroom is intimidating or patronizing. The positioning implies that the knowledge is being transferred in only one way—from teacher to student. Successful adult education attempts to minimize this perception by making the interaction more equal. This can be done by accounting for the participants' practical knowledge and by discussing what the participants know about the topic at the beginning of an activity. Doing so creates an exchange of knowledge and recognizes the contributions of each participant to the process, thereby facilitating the process of engagement, and enables the knowledge transfer professional to adapt the activities to the learners based on what was said. This enriches the learning experience for all.

"Train-the-trainer" or "near peer" education is another way to transfer knowledge in an environment acceptable to adult learners. The Oregon State University Cooperative Extension Service's Master Gardener and Master Woodland Manager programs are examples of this practice. Participants are taught in-depth, comprehensive information with the expectation that they will subsequently share their new knowledge with their peers. Currently, no such formalized programs appear to exist in landscape ecology.

8.5.1.3. Incorporate a Range of Teaching Modalities to Accommodate Different Learning Styles

People perceive and process information in different ways. Understanding these differences and incorporating methodologies that account for these different learning styles enhances knowledge transfer. In North America, lectures remain the dominant teaching method, even though it is believed that approximately half of the population has difficulty processing oral information (Simon-Brown 1999). Learning styles are defined as a biologically and developmentally imposed set of personal characteristics that make a given teaching method more effective for some than for others (Dunn et al. 1989; Dunn and Griggs 1988, cited in Reeb 2003). Learning *style* influences how a person learns best and should not be confused with the *ability* to learn. Various social scientists have developed models that propose auditory, visual, kinesthetic, and tactile learning modalities (Simon-Brown 1999).

To overcome the problem of different preferred styles and to stimulate retention, knowledge transfer professionals typically incorporate a variety of techniques in each session. These include hands-on activities; varying amounts of individual, small-group, and full-group work; practicing active listening; offering both practical and conceptual information; conducting field trips; encouraging journal-keeping and role-playing; and creating problem-solving teams. Most students take advantage of most or all learning modalities during learning, but to a different extent for each student and each modality (Reiff 1992, cited in Reeb 2003). An individual's dominant modality offers them the most efficient processing of new information, especially when the person is fatigued or under stress. However, taking advantage of a person's secondary modality enhances, clarifies, or supplements the dominant one, without interfering with it (Wislock 1993, cited in Reeb 2003).

8.5.1.4. Adapt to the Independent, Self-Directed Nature of Adult Learners

Most adult learners are self-directed problem-solvers. Such individuals may only be willing to learn concepts if it is clear to them that the concepts are a necessary step toward solving a problem that is important to them. Learning information for its own sake is not the norm; the information must be meaningful to them in their present situation. Adult learners will typically seek information when they need it, and should be encouraged to take charge of their own learning (Wise and Ezell 2003). Motivated adult learners incorporate new information into what they already know to develop action-oriented solutions.

Electronic technologies can enhance the knowledge transfer professional's ability to accommodate the needs of these "just in time" learners. The tools of the trade can be synchronous or asynchronous. *Synchronous* (simultaneous) tools include teleconferencing, instant messaging, real-time Web work, and satellite linkages. These approaches offer the advantage of interactions between student and teacher and among the students, but may force students to participate at inconvenient times. *Asynchronous* (not simultaneous) approaches include non-real-time Web work, streaming video, e-mail discussion groups, virtual field trips, videoconferencing, cable TV, CDs, and DVDs, and can also be effective if they are designed according to the best practices for successful instructional design. These approaches offer the advantage of flexible training that can be delivered at the time most suitable for the student, but may reduce the ability for participants to interact.

8.5.1.5. Adopt a Minimalist Philosophy

One of the most difficult techniques for knowledge transfer professionals to master is referred to as minimalism (Carroll 1998), also called the "less is more" philosophy. It is more useful for learners to cover less information and to explore the meaning of that information than it is to rush them through a large amount of material (Norland 2003). Teaching the major concepts and then providing students with the "how to" tools that allow them to locate the specific information they need is one way to overcome the need to "tell all." Providing a myriad of optional in-depth background materials to support the primary knowledge transfer is another way to minimize nonessential instruction. Modeling is a third way in which knowledge transfer professionals can demonstrate the planning, organizational, and decisionmaking strategies that are being taught.

8.5.1.6. Document Personal and Group Achievements

Effective knowledge transfer acknowledges, formally and informally, the achievement of certain learning milestones. Documenting that learning has occurred by awarding certificates or credentials works well because these proofs of accomplishment are valued both within the educational setting and by the larger external community. Participants acquire a sense of accomplishment, but credentials also reinforce the value of and legitimize the knowledge transfer experience.

8.5.2. Credible Research-Based Information

Accurately communicating research information in ways that meet audience needs, without changing the fundamental nature of the information, is the crux of successful knowledge transfer. To maintain our legitimacy and the trust of our audience, knowledge transfer professionals must scrutinize the materials we produce to ensure their credibility. Materials that have been formally and rigorously reviewed by peers, referees, and panels of scientists can generally be recommended. Scanning the research literature is a good way to subject research to a reality check: it provides a better grasp of the contextual framework for the issue (Adams and Hairston 1994) and insights into how well proven or accepted a conclusion may be. Moreover, maintaining a broad familiarity with research in a variety of fields allows the knowledge transfer specialist to integrate a wider spectrum of research results and provide a more holistic understanding of the subject matter (Krueger and Kelley 2000).

Credible research-based knowledge transfer activities are not prescriptive. Rather than telling the audience what to do, they offer a continuum of alternatives and a discussion of the consequences of each alternative; in addition, they provide diagnostic or decision-support tools that let the audience identify the advantages and drawbacks of each alternative, and make wiser decisions on this basis. The activities do not endorse one practice over another, but rather leave this decision to the audience. This key characteristic distinguishes education from advocacy.

Garland (1997) states that participants should be able to trust the information they receive and act on it to:

- learn about the available options and their consequences
- identify the relevant facts for each option
- distinguish among values, myths, opinions, and facts
- identify any personal values that are involved
- identify unknowns and variables
- use data to analyze individual situations
- define what success would look like for them

All of these principles should be incorporated into planned transfer activities.

8.5.3. Evaluation

Evaluation has various definitions. In this chapter, we use the term to describe the systematic collection, analysis, and reporting of information that can be used to improve programs or activities. Evaluation is also a continuous process of inquiry— a process of asking questions about the social, economic, and environmental conditions and circumstances within which knowledge transfer occurs. Evaluation helps to answer the following questions:

- Are my knowledge transfer efforts making a difference?
- What changes would make my efforts more effective?
- How can I refine future activities to achieve better results?
- To what extent is my audience using the information?

Knowledge transfer specialists can clearly benefit from a rigorous evaluation that answers these questions. In addition to improving future activities, the evaluation

Fundamentals of Knowledge Transfer and Extension

data can raise more questions, which in turn lead to more research. The evaluation can also provide justification for additional research funding. From a management viewpoint, the results of the evaluation can be the basis for performance appraisals and for garnering support from stakeholders, including legislators and granting agencies.

A quick Web search will reveal dozens of methods for evaluating the effectiveness of knowledge transfer. Two standard tools for planning and assessing impacts are "logic models" and various versions of Bennett's hierarchy. A logic model visually displays the sequence of actions that describe what the activity is and what it will accomplish (Kellogg Foundation 2000). It is particularly effective in the natural resources arena since it directly links the problems (situations) to the interventions (inputs and outputs) and to the impacts (outcomes) (McCawley 2001). It illustrates the connections between available resources, activities carried out with audiences, the services delivered, and the intended results, as well as the long-term goal to which the activity contributes. As an evaluation tool, it helps to identify process and outcome indicators, highlight elements that will yield useful evaluation data, and select an appropriate sequence for collecting data and measuring progress (McCawley 2001).

Bennett's Hierarchy of Evidence (Bennett 1975) describes a series of staircase levels of evidence of program impacts (Figure 8.3). Beginning at the bottom step with inputs and progressing upwards to the end result, evidence of program impact

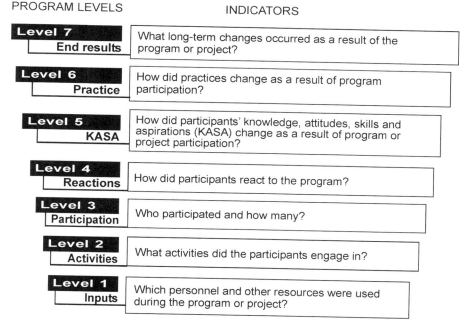

Figure 8.3. Bennett's hierarchy of evidence (Bennett 1975) provides a system for evaluating the impacts of transfer programs. Evaluation becomes progressively more reliable but also more costly (from levels 1 to 7).

at each ascending step is progressively more substantial and more reliable for decisionmaking, albeit more difficult, costly, and time-consuming to measure (Suvedi and Morford 2003).

A newer version of Bennett's Hierarchy, called Targeting Outcomes of Programs (TOP), includes a downward staircase that targets outcomes, tracks progress toward achieving these outcomes, and helps managers evaluate the degree to which activities affect the targeted social, economic, and environmental conditions (Bennett and Rockwell 1995).

Applying any of these methods following transfer activities helps to evaluate and to improve future knowledge transfer efforts.

8.6. SUMMARY AND ONGOING CHALLENGES

The various processes and approaches to knowledge transfer that have been described in this chapter are proven methods to successfully transfer knowledge to an audience. Additional considerations and ongoing challenges for transfer professionals and forest landscape ecologists seeking knowledge transfer success include the following:

- Strive to identify and engage new audiences, while building increasingly strong relationships with existing audiences.
- Seek new and emerging communication technologies that will let us match our approaches to each audience member's needs and abilities. Monitor these options, and strive to provide a diverse mix of approaches to reach more people, more effectively.
- Recognize that communication within and between organizations remains a challenge. Successful knowledge transfer organizations as well as individuals continue to build an increasing sense of community and teamwork.
- Recognize that learning is an ongoing activity throughout a professional's career, and that knowledge transfer activities are not complete just because a program is complete. Knowledge transfer will continue as new information and research results become available, and as feedback from audiences identifies problems with existing knowledge and new needs. Successful learning requires continuous engagement of the learner, and a two-way exchange of knowledge between the transfer specialist and the learner.
- Seek innovative sources of funding to ensure that important long-term programs can continue, while still providing the flexibility to fund short-term programs or activities that respond to sudden changes in conditions. In some cases, a user-pay model may be appropriate, particularly where this approach pays for the cost of an activity that might otherwise go unfunded.
- Evaluate transfer efforts, perhaps by investigating more rigorous ways to document the return on investment from an activity or program. Evaluations should focus on outcomes, and outcomes should be considered in three areas

(OSU 2004): progress toward achieving a program's goals, the benefits for the public good, and the benefits for the audience that has received the transferred knowledge.
- Since circumstances change, plan to periodically assess the situation and, if necessary, redirect efforts and resources. Planning should be both strategic, to cope with long-term situations, and tactical, to cope with short-term or sudden crises.
- Move beyond simple cooperation by striving for collaboration and, eventually, for full partnership.

LITERATURE CITED

Adams PW, Hairston A (1994) Using scientific input in policy and decision making. Oregon State Univ, Extension Serv, Corvallis. EC 1441.

Bennett C (1975) Up the hierarchy. J Extension 13(2):7–12. (http://www.joe.org/joe/1975march/1975-2-a1.pdf)

Bennett C, Rockwell K (1995) Targeting outcomes of programs (TOP): an integrated approach to planning and evaluation. Univ Nebraska, Lincoln. (http://citnews.unl.edu/TOP/english/overviewf.html)

Bergstrom A, Clark R, Hogue T, Iyechad T, Miller J, Mullen S, Perkins D, Rowe E, Russell J, Slinski M, Snider B, Thurston F, Simon-Brown V (eds) (1996) Collaboration framework—addressing community capacity. Ohio State Univ, Columbus. USDA Coop State Res Educ Extension Serv, National Network for Collaboration. (http://crs.uvm.edu/nnco/collab/framework.html)

Boyer EL (1990) Scholarship reconsidered: priorities of the professoriate. Carnegie Foundation for the Advancement of Teaching, Princeton Univ Press, Princeton.

Carroll J (ed) (1998) Minimalism beyond the Nurnberg funnel. MIT Press, Cambridge, Massachusetts. 350 pp.

Cooper A, Graham D (2001) Competencies needed to be successful county agents and county supervisors. J Extension 39(1). (http://www.joe.org/joe/2001february/rb3.html)

CRC (1998) Active listening. Conflict Research Consortium. (http://www.colorado.edu/conflict/peace/treatment/activel.htm)

DeYoe D, Hollstedt D (2003) A knowledge exchange system: putting innovation to work. Paper presented at the 6th IUFRO Extension Working Party Symposium, Troutdale OR. 8 pp.

Dunn R, Griggs S (1988) Learning styles: quiet revolution in American secondary schools. National Association of Secondary School Principals, Reston.

Dunn R, Beaudry J, Klavas A (1989) Survey of research on learning styles. Educational Leadership 46(6):50–58.

Fletcher R (1999) Forestry extension via community based university faculty. Paper presented at the 4th IUFRO Extension Working Party Symposium, Bled, Slovenia. 9 pp.

Garland JJ (1997) The players in public policy. J For 95(1):13–15.

Gustafson E, Zollner P, Sturtevant B (2004) Influence of forest management alternatives and land type on susceptibility to fire in northern Wisconsin, USA. Landscape Ecol 19(3):327–341.

Hogue T, Miller J (2000) Effective collaboration: strategies for pursuing common goals. Rocky Mountain Press, Longmont.

Johnson J (2003) Best practices in forestry extension—a state perspective. Paper presented at the 6th IUFRO Extension Working Party Symposium, Troutdale, OR. 9 pp.

Kellogg Commission (1998) Returning to our roots: the engaged institution. Kellogg Commission on the Future of State and Land-Grant Universities, National Association of State Universities and Land-Grant Colleges, Washington.

Kellogg Foundation (2000) Logic model development guide: using logic models to bring together planning, evaluation, and action. (http://www.wkkf.org/Pubs/Tools/Evaluation/Pub3669.pdf)

Krueger W, Kelley C (2000) Describing and categorizing natural resources literature. Oregon State Agric Exp Stn, Corvallis. Tech Pap Number 11661.

McCawley PF (2001) The logic model for program planning and evaluation. University of Idaho Extension, Moscow. CIS 1097.

Norland E (2003) Best practices in extension forestry: what we know and what we do. Paper presented at the 6th IUFRO Extension Working Party Symposium, Troutdale, OR. 9 pp.

OSU (2002) Faculty handbook. Oregon State Univ, Corvallis. (http://oregonstate.edu/facultystaff/handbook/)

OSU (2004) OSU extension strategic plan: a framework for engaging Oregonians. Oregon State Univ, Extension Serv, Corvallis.

Patterson T Jr (1991) Tomorrow's extension educator—learner, communicator, systemicist. J Extension 29(1). (http://www.joe.org/joe/1991spring/fut1.html)

Reeb J (2003) Taking into account different learning modalities when teaching adults complex subject matter: an example for improving manufacturing. Paper presented at the 6th IUFRO Extension Working Party Symposium, Troutdale, OR. 12 pp.

Reed AS (1999) Key factors in the success of extension forestry programs. Paper presented at the 4th IUFRO Extension Working Party Symposium, Bled, Slovenia. 6 pp.

Reichenbach M, Simon-Brown V (2002) Linking strategic thinking and project planning: the Oregon State University extension forestry experience. J Extension 40(4). (http://www.joe.org/joe/2002august/iw1.shtml).

Reiff J (1992) What research says to the teacher: learning styles. National Education Assoc, Washington.

Simon-Brown V (1999) Effective meetings management. Oregon State Univ, Extension Serv, Corvallis. Working together series. EC 1508.

Suvedi M, Morford S (2003) Conducting program and project evaluations: a primer for natural resource program managers in British Columbia. FORREX-Forest Research Extension Partnership. Kamloops. FORREX Series 6.

Wise D, Ezell P (2003) Characteristics of effective training: developing a model to motivate action. J Extension 41(2). (http://www.joe.org/joe/2003april/a5.shtml)

Wislock R (1993) What are perceptual modalities and how do they contribute to learning? New Directions for Adult and Continuing Education 59:5–15.

9

Synthesis: What Are the Lessons for Landscape Ecologists?

Thomas R. Crow, Ajith H. Perera, and Lisa J. Buse

9.1. Lessons from the Book
    9.1.1. Knowledge Transfer Is Necessary
    9.1.2. Knowledge Transfer Is an Active Process
    9.1.3. Knowledge Transfer Experiences Are Diverse
    9.1.4. Knowledge Transfer Benefits Developers
9.2. Where Do Knowledge Developers Go from Here?
    Literature cited

9.1. LESSONS FROM THE BOOK

The main goal of this book was to create an awareness of the need for knowledge transfer among forest landscape ecologists. To that end, we considered aspects of knowledge transfer and extension in general, critically examined the aspects of transfer that are unique to forest landscape ecology, and highlighted several examples of successful landscape ecological knowledge transfer. In the preceding chapters, we have explored various facets of the application of landscape ecology in forest policy and management from a North American perspective. In this chapter, we summarize the main messages contained in the book.

THOMAS R. CROW • USDA Forest Service, Research and Development, Environmental Sciences, 1601 N. Kent Street, Arlington, VA 22209, USA. AJITH H. PERERA and LISA J. BUSE • Ontario Ministry of Natural Resources, Ontario Forest Research Institute, 1235 Queen St. E., Sault Ste. Marie, ON P6A 2E5, Canada.

9.1.1. Knowledge Transfer Is Necessary

A considerable gap is evident between the large body of forest landscape ecological knowledge and its application. This gap exists and may continue to widen even as the potential for applications expands because the flow of knowledge from developers to users is not automatic. Many factors favor the application of forest landscape ecological knowledge. Users of landscape ecological knowledge are many and range from legislators to forest policymakers, planners, and managers, with each group having unique information needs at different scales (as emphasized by Buse and Perera 2006; King and Perera 2006; Perera et al 2006). As forest managers begin to consider larger scales, the potential for application of this knowledge is also expanding. Computing technology, once considered an obstacle, has advanced and become more accessible; combined with readily available and relatively inexpensive data, this technological capacity is now less of an impediment to applying landscape ecological knowledge. However, other barriers to knowledge flow to users still exist, such as a lack of awareness of the available knowledge, the complexity and unfamiliarity of the knowledge, and the fact that much landscape ecological knowledge is not available in a directly usable form.

Much of the unfamiliarity stems from the breadth of the spatial and temporal scales that define landscape ecology (King and Perera 2006). Given the infeasibility of typical cause-and-effect experimentation at broad scales, simulating scenarios and exploring if–then situations using simulation models have become the research tools of choice in landscape ecology. Simulation models are not only a principal vehicle for experimentation and generation of knowledge, but are also useful to transfer knowledge. They may be unpalatable to potential users for many reasons: unfamiliarity with the technology; lack of understanding of the purpose of the model; unclear assumptions; discomfort with abstract concepts, coarseness of the model's scale, stochasticity, and complexity; and distrust of the mechanisms underlying the model and conceptual validation methods. As Gustafson et al. (2006) indicated, user difficulties with models can lead to inappropriate use and ultimately to rejection of the models. These can be avoided by proactive knowledge transfer.

9.1.2. Knowledge Transfer Is an Active Process

Developers of landscape ecological knowledge should actively partake in transferring knowledge to potential users. Several broad categories of approach can help developers accomplish this transfer: supply-driven ("push"), demand-driven ("pull"), and collaborative-iterative processes, as well as various combinations of the three, can all be potentially useful depending on the nature of the audience, the stage of development of the application, and the nature of the knowledge transferred (Perera et al. 2006). Regardless of the approach, the applications, and the users, landscape ecologists must first understand the fundamentals of knowledge transfer.

Reed and Simon-Brown (2006) describe in detail some key considerations for the developers of landscape ecology knowledge who wish to engage in knowledge transfer, many of which are illustrated in practice in the case study chapters (Buse and Perera 2006, Gustafson et al. 2006; Hampton et al. 2006; Lytle et al 2006).

All case studies stress that the first step is to identify the primary users of the knowledge and engage them from the outset of the knowledge development process rather than waiting until after the knowledge has been developed. Early engagement helps knowledge developers to identify specific user needs and develop a working rapport with users that persuades this audience their needs are being met and that their input is valued. Flexibility and objectivity in the approach to selecting and implementing specific knowledge transfer mechanisms are important because users differ in their learning styles. Transferring concepts (knowledge) first and allowing users to explore further and apply their knowledge by providing access to appropriate tools (technology) is an effective means for users to discover alternatives rather than relying on knowledge developers to provide a single, possibly suboptimal, solution. This approach reinforces the landscape ecological concepts, strengthens the relationship between the developers and users of knowledge, and supports a process of continuous engagement.

As well, transfer is an interactive process, in which both developers and users benefit from continuous engagement. It enables knowledge developers to be flexible so as to adapt their approach to the needs of the users, and users to become progressively comfortable with the new knowledge or tools in incremental stages. Participants in knowledge transfer must clearly understand the specific needs and characteristics of the users, whether that knowledge informs policy or becomes a management tool. The relationship between users and developers must be collaborative and is best established early and fostered continually.

Ultimately, the goal of the transfer process is to elevate the level of engagement from cooperation to collaboration and eventually to an ongoing partnership (Reed and Simon-Brown 2006). As described by Perera et al. (2006), the details of the knowledge transfer process may be complex, but the overall process can be conceptualized simply as a flow of information among developers (e.g., researchers), practitioners (e.g., users), and transfer specialists (e.g., extension and GIS specialists) by means of ongoing engagement and communication. The knowledge transferred through this process can range from conceptual principles to user tools to data. Although these fundamentals are broadly applicable, the specific techniques and approaches required may vary depending on the knowledge being transferred, the nature of the audience, and the stage in the knowledge transfer process.

9.1.3. Knowledge Transfer Experiences Are Diverse

The case studies of applications of landscape ecology in forestry presented in this book range from experience with transferring a single user tool to one user group in a forest management area to experience transferring a variety of concepts, knowledge, and user tools to a hierarchy of diverse user groups in a national forest management agency. Despite their geographical, cultural, and situational differences, many commonalities are evident among these case studies, particularly in how knowledge developers approached the transfer process and what they considered important to a successful outcome (Table 9.1). For example, knowledge developers who strive to

Table 9.1. A synoptic comparison of knowledge transfer experiences reported in case study chapters, with the goals and audiences broadening from Gustafson et al. (2006) to Crow (2006)

	Case study chapters				
	Gustafson et al. (2006)	Hampton et al. (2006)	Lytle et al. (2006)	Buse and Perera (2006)	Crow (2006)
Setting	A forest management area within one category of land ownership	A large forest management area with multiple land ownerships	A forested region with multiple land ownerships across the border between Canada and the U.S.	A subnational (provincial) forest management agency	A national forest management agency
Audience	Forest managers	Forest managers, the public, specific stakeholders	Forest managers in public land management agencies, and specific stakeholders	A hierarchy of decisionmakers in a public land management agency, forest managers, and stakeholders	
Transfer goals	Transfer technology to support forest management planning	Transfer knowledge and tools for forest assessment and planning	Transfer knowledge and technology to support strategic forest planning	Incorporate concepts, knowledge, and technology in the development of legislation and policy, in strategic planning, and in forest management	
Material transferred	Models, tools	Concepts, data, tools	Knowledge, models, tools	Concepts, knowledge, models, data, tools	
Participants	Developers, users, local experts	Developers, users, the public, stakeholders	Developers, users, public agencies, stakeholders	Developers, transfer specialists, local experts, users, GIS technologists	Developers, managers, planners, policy-makers, stakeholders
Transfer approaches and methods	Continuous personal interaction between developers and users through workshops, discussion forums, and informal communication Collaborative, iterative approach	Push-based and collaborative approaches	Push-based, pull-based, and collaborative approaches	A variety of methods, including push-based, pull-based, and collaborative approaches	

Keys to success	Shared vision of the outcome and commitment of all participants			Political will to adapt, local capacity to generate knowledge, and an enabling technological and personnel infrastructure	Demonstrating an ability to solve critical issues
Major challenges	Technological barriers, and the lack of a common language	Organizational cultures, resource constraints, and technological barriers	Land ownership complexity, difficulties coordinating among numerous and dispersed partners, shifting organizational priorities	Complexities associated with multiple, diverse audiences; shifting organizational priorities; and bureaucratic barriers	Institutional and organizational barriers, and an inadequate technological infrastructure

transfer their findings to users should be aware that it requires a long-term and continuous commitment of both time and resources. The lack of a common language can slow the transfer process initially. As well, it can be difficult to empower audiences without oversimplifying the issues. Use of a common language helps to establish a common vision, goals, and commitment. As well, the case studies clearly reveal the value of establishing the context for the knowledge transfer, and emphasize the transfer of concepts first, even when the transfer of tools is the final goal.

The case studies also revealed challenges to knowledge transfer in landscape ecology. In contrast to the above-mentioned commonalities in the success factors, challenges are more difficult to generalize because they tend to depend on the situation. The most commonly cited challenge relates to institutional barriers stemming from the diverse organizational structures and cultures of stakeholder and knowledge developer organizations. Technological barriers, although diminishing, remain in some instances.

As Gustafson et al. (2006) suggest, engaging in a collaborative, iterative approach in transferring knowledge and user tools is effective when users and developers are equally committed and share a common desired outcome. Engaging local experts as partners in addition to users and developers can improve the efficacy of the process. The collaborative-iterative approach is preferred to push-only (developer initiated) or pull-only (user initiated) approaches when a specific tool will be transferred to a particular user group to accomplish a specific purpose. The time, effort, and commitment required from all participants may preclude exclusive use of this approach when the transfer material, application goals, and audience are more complex. This is evident in the experiences of Hampton et al. (2006) and Lytle et al. (2006), for which the user audiences were diverse and the transfer goals were broad: Because of the intense time commitments that arise from the long time frame often associated with complex transfer situations, push and pull approaches complemented the collaborative-iterative approach and helped to establish effective relationships. Lytle et al. pointed out the importance of identifying and engaging leaders within each of the intended user organizations to champion the process. In addition, adopting a flexible approach by resorting to a suite of transfer methods is beneficial. Hampton et al. (2006) emphasized the advantages of using transfer to support decisionmaking rather than to generate or advocate solutions. At this scale, differences in organizational cultures begin to affect knowledge transfer, and the relative effort spent on building and maintaining relationships and providing opportunities for engagement among users becomes significant.

Evidence of knowledge transfer at the institutional scale is present in policies, strategic plans, and management practices at both a national scale (Crow 2006) and a subnational scale (Buse and Perera 2006). Although it is difficult to generalize the suitability of specific transfer techniques in these instances, it is apparent that a combination of push-based, pull-based, and collaborative-iterative approaches are relevant. The presence of an institutional will to adapt is the major reason for success in knowledge transfer at broad scales. The major challenges are also institutional, and include bureaucratic barriers, shifting priorities, and political realities (Buse and Perera 2006). In addition, the composition of the audience and stakeholders becomes

What Are the Lessons for Landscape Ecologists?

extremely diverse and complex. Experiences at the institutional scale also suggest that the integration of landscape ecology knowledge into policies and strategic plans (and even into legislation) as a result of knowledge transfer is possible, but that this requires a sustained effort over a longer period, and the use of more than one approach.

9.1.4. Knowledge Transfer Benefits Developers

Transfer of landscape ecological knowledge should not be seen as a process that only benefits the users; as Gustafson et al. (2006) and King and Perera (2006) suggest, there are also many advantages for knowledge developers. One is that the transfer process offers a form of peer review in which the users of knowledge provide feedback on its *applicability*; this is clearly different from peer review by colleagues, which focuses only on the scientific content, often irrespective of its practical relevance. This review not only improves the final application of the knowledge but also increases confidence in its use. The collaborative-iterative approach is an excellent example of peer review and feedback that progressively enhances the quality and applicability of the knowledge and leads to shared ownership of the transferred knowledge. Communication between developers and users during the transfer process also provides opportunities for developers to gain valuable insights that might not be available through customary discussions with their peers. Such insights can provide important guidance for future research efforts. As well, ongoing dialogue with potential users of landscape ecological knowledge helps to broaden the developer's perspective and, in academic environments, may expose graduate students in forest landscape ecology to real-world scenarios that help them appreciate the potential for application of their knowledge. Finally, forest landscape ecology is an applied science in which research knowledge is developed specifically for use in forest management. Engaging in knowledge transfer provides developers with an opportunity to view the benefits of their research efforts.

However, as Perera et al. (2006) point out, successful applications should not be confused with successful transfer. Although the ultimate success of transfer is reflected in advances in the application of knowledge, this is not the sole determinant of a successful transfer process. For example, transfer can be deemed successful when users understand the concepts, use the tools appropriately, and can apply the knowledge they have gained. Application of that knowledge in developing policies or practices may not occur because successful implementation results from myriad other influences unrelated to the knowledge exchange between developers and users of the knowledge.

9.2. WHERE DO KNOWLEDGE DEVELOPERS GO FROM HERE?

As we learned, the transfer of forest landscape ecological knowledge is possible under a range of scenarios, from implementation of a single tool that will influence a limited set of decisions to the development of policies with a broad range of social repercussions. When the transfer situation becomes more complex, from single to multiple applications, one to many user groups, single to multiple organizations, one to many ownerships, and narrow to wide impact of the application, common keys to success as well as challenges emerge. In addition, no single transfer method or list of obstacles to be overcome can be identified before engagement between knowledge developers and users begins because each situation is unique. There are also many participants in knowledge transfer beyond developers and users, such as transfer specialists and other experts, and all of them share partial responsibility for the process. Amidst these complexities, researchers must accept the responsibility to identify the needs and opportunities for application of their knowledge and to ensure transfer of the knowledge they develop.

Imagine the following scenario. A group of elected officials visits a forestry research agency. The officials are well aware that the agency's scientists conduct outstanding research and that their work and the publications resulting from their research are held in high regard by the broader scientific community. But the officials are not interested in exemplary publications produced by renowned scientists; instead, they want to know about the relevance of the work, how it could solve important problems, whether the researchers accomplished their original goals, and—not surprising given that these are elected officials—whether the work will help their constituents. Not only do the scientists need to make clear the relevance of their research but they also have to present their science in a manner that makes sense to the elected officials. Furthermore, the scientists have only a few minutes to make their case before the policymakers hurry off to their next appointment.

Although this scenario is purely hypothetical, researchers who receive government funding will recognize its plausibility. Those responsible for funding scientific research increasingly want to know what they are getting for their money, and want to receive this information in clear and unambiguous terms. They want to know about outcomes, not just outputs. Unfortunately, though scientists are trained to communicate with their peers, there is much less emphasis placed on communicating with the much larger and more diverse audience of policymakers, knowledge users (such as planners and managers), public officials, and the general public. As Scheuering and Barbour (2004) observed, "Science does not exist in a vacuum, but reading scientific publications might make you think it does."

During these times of decreasing funding for research and increasing accountability of researchers to those who fund their work, the need to close the gap between those who produce knowledge and those who use it is growing. As we have stressed in this book, this requires a reciprocal relationship in which a partnership is formed; in the case of forest science, the partnership is between those who manage the natural resource and those who study the resource, and the partnership exists for their mutual benefit. Although the importance of this relationship between producer and consumer of knowledge has been stated many times before, it is worth repeating. Bridging this gap calls for fundamental changes in the ways that universities train both the producers and the consumers of knowledge and it requires changes in the ways research organizations reward their scientists. In an interesting essay on the role of the university, Rowe (1990) argued that universities have become "overloaded and top-heavy with expertness and information." Instead of being "a know-how institution" they should become "know-why institutions." The know-how approach is rich with information but poor in knowledge. It is this knowledge and the basic understanding that provides "ethical alternatives on which to act." As a profession, we researchers are good at collecting information; we also need to turn this information into knowledge that is useful to those who support our efforts.

In making our case for knowledge transfer, we also must recognize the pitfalls. Many of these have been identified in the preceding chapters. One, however, deserves special attention. If research is justified solely on its perceived merits to society, there is a risk of failing to support programs that are presently "out of favor" but that nonetheless have value, as well as high-risk ventures that constitute some of the research community's most innovative work. We contend, however, that by closing the gap between producers and consumers of knowledge, the likelihood of support for this research is increased, not diminished; people will support what they understand more readily than abstract concepts that appear to have no relevance. This is also true of funding agencies: research funds will be more readily awarded when the agency understands how the research helps meet the agency's goals.

Those involved in landscape ecology, and specifically in forest landscape ecology, have been successful in persuading the policy community that our science should be taken seriously (Klijn 2005). A landscape perspective, with its emphasis on spatial relationships, on collaboration across disciplines, on multiple scales and hierarchies, and on the importance of context and local processes, is the right science at the right time for resource managers. Consequently, the most important job for researchers is to ensure that this science does not operate in a vacuum, and to act on opportunities for the application of landscape ecological knowledge. We hope that by introducing the concept of knowledge transfer to the vocabulary of forest landscape ecological researchers, this book will serve as a catalyst for future endeavors to improve the effectiveness of knowledge transfer and will contribute to successful application of this knowledge.

LITERATURE CITED

Buse LJ, Perera AH (2006) Applications of forest landscape ecology and the role of knowledge transfer in a public land management agency. In: Perera AH, Buse LJ, Crow TR (eds) Forest landscape ecology: transferring knowledge to practice. Springer, New York.

Crow TR (2006) Moving to the big picture: applying knowledge from landscape ecology to managing U.S. National Forests. In: Perera AH, Buse LJ, Crow TR (eds) Forest landscape ecology: transferring knowledge to practice. Springer, New York.

Gustafson EJ, Sturtevant BR, Fall A (2006) A collaborative iterative approach to transferring modeling technology to land managers. In: Perera AH, Buse LJ, Crow TR (eds) Forest landscape ecology: transferring knowledge to practice. Springer, New York.

Hampton HM, Aumack EN, Prather JW, Dickson BG, Xu Y, Sisk TD (2006) Development and transfer of spatial tools based on landscape ecological principles: supporting public participation in forest restoration planning in the southwestern United States. In: Perera AH, Buse LJ, Crow TR (eds) Forest landscape ecology: transferring knowledge to practice. Springer, New York.

King AW, Perera AH (2006) Transfer and extension of forest landscape ecology: a matter of models and scale. In: Perera, AH, Buse LJ, Crow TR (eds) Forest landscape ecology: transferring knowledge to practice. Springer, New York.

Klijn F (2005) Landscape ecology as the broker between information supply and management application. In: Wiens J, Moss M (eds) Issues and perspectives in landscape ecology. Cambridge Univ Press, Cambridge, UK, pp 181–192.

Lytle DE, Cornett MW, Harkness MS (2006) Transferring landscape ecological knowledge in a multi-partner landscape: the Border Lakes region of Minnesota and Ontario. In: Perera AH, Buse LJ, Crow TR (eds) Forest landscape ecology: transferring knowledge to practice. Springer, New York.

Perera AH, Buse LJ, Crow TR (2006) Knowledge transfer in forest landscape ecology: a primer. In Perera AH, Buse LJ, Crow TR (eds) Forest landscape ecology: transferring knowledge to practice. Springer, New York.

Reed AS, Simon-Brown V (2006) Fundamentals of knowledge transfer and extension. In: Perera AH, Buse LJ, Crow TR (eds) Forest landscape ecology: transferring knowledge to practice. Springer, New York.

Rowe S (1990) Home place. Essays on ecology. NeWest Publishers Ltd, Edmonton, Alberta.

Scheuering R, Barbour J (2004) Focused science delivery makes science make sense. Western Forester 49(5):1–3.

Index

Page numbers followed by f and t indicate figures and tables, respectively

A

Accessibility, for knowledge transfer, 186
Acts
   Crown Forest Sustainability Act, 6, 133–135
   Forest and Rangeland Renewable Resources Planning Act (RPA), 166
   Healthy Forest Restoration Act, 74, 83, 109–110, 160, 162
   National Environmental Policy Act, 79, 86, 107
   National Forest Management Act, 79, 158
   Wilderness Act, 105
Atlantic white cedar (*Chamaecyparis thyoides*), 108, 167
Augusta Creek Study, 173–174

B

Balsam fir (*Abies balsamea*), 101, 108, 167
Beech (*Fagus* spp.), 169
Bennett's Hierarchy of Evidence, 201–202
BFOLDS, *see* Boreal Forest Landscape Dynamics Simulator
Black spruce (*Picea mariana*), 98, 101–102, 108
Border Lakes landscapes, in Minnesota and Ontario, 98, 100
   challenges to achieving desired future condition in, 110–111
   ecology of, 101–102
   land ownership and management goals in, 102–104
   landscape description of, 99–100
Border Lakes landscapes in Minnesota and Ontario, fundamental rationales for
   overlapping stakeholder missions, 104–106
   scale of ecological processes and land ownership patterns, 107
Border Lakes Partnership, 98
   goal of, 99
Boreal Forest Landscape Dynamics Simulator (BFOLDS), 143–144

C

Coastal Landscape Analysis and Modeling Study (CLAMS), 172
Coconino National Forest, 89
Collaborative process, 77
Collaborative relationship, 189
Community Wildfire Protection Planning process, 82, 85
Cooperative relationships, 188–189
Coordination, for knowledge transfer, 186–187
Credible research-based information, for knowledge transfer, 199–200
Crown Forest Sustainability Act, 133–136

215

D

Decision makers, 46
  hierarchy of, 8
Decision matrix, 75–76
Decision-support models, 45

E

Ecoregions, 5–6
Ecosystems, broad-scale, 2
Engagement in knowledge transfer, factors for enhancing, 184
  accessibility, 186
  coordination, 186–187
  integration, 186
  neutrality, 185–186
  resource partnerships, 187
  respect for audience, 185
  responsiveness for, 185
Evaluation, for knowledge transfer, 200–202
Extension specialists, 12

F

Fire Learning Network, 126
Forest and Rangeland Renewable Resources Planning Act (RPA) of 1974, 166
Forest Ecosystem Restoration Analysis (ForestERA)
  data and tools, 85–86
  knowledge transfer, successes and challenges in, 85–92
  knowledge transfer by, 70–80
  knowledge transfer mechanisms, 80–85
  objectives, 67–70
  spatial decision-support system, 72
  tools for, 78, 87–88
Forest ecosystem succession models, 99
ForestERA, *see* Forest Ecosystem Restoration Analysis
Forest fragmentation, 170
Forest landscape, aspects of, 4
  disturbance, 4
  forest landscape change, 4
  forest landscape function, 4
  forest management and planning strategies, 4
  habitat provision, 4
  spatial heterogeneity, 4
Forest landscape and regional change analyses, 168–171
  factors contribute to, 169
Forest landscape ecological knowledge, causes of unavailability
  incompatibility with existing infrastructure, 11
  incompatibility with their needs, 11
  lack of awareness, 11
Forest landscape ecological knowledge, difference of
  breadth of spatial scale, 10
  focus on coarser resolution, 11
  length of temporal scale, 11
  multidisciplinary complexity, 10
  reliance on conceptual models, 11
  stochasticity in broad landscape processes, 11
Forest landscape ecological knowledge transfer, *see* Knowledge transfer
Forest landscape ecological knowledge users
  decision makers, 140
  forest resource managers, 141
  policy makers, 140–141
Forest landscape ecological simulation models
  barriers to application of, 24–30
  components of, 22
  for decision making, 25
  definitions, 21–23
  generic barriers, 21
  mutual benefits, 39–40
  users, 23–24
Forest landscape ecological simulation models, barriers to application of
  assumptions and application limitations, 27
  dissatisfaction with abstractions and assumptions, 27
  distrust of "Black Box" models, 29
  distrust of methods of model validation, 29–30
  modeling at large scales, discomfort with, 28
  necessity for third party involvement, 30

Index

stochasticity and variability in simulated processes, problems with, 28–29
unavailability of computing technology and spatial data, 30
uncertainty about purpose of models, 26
unfamiliarity with topic, 26
Forest landscape ecologists
  knowledge transfer in forest landscape ecology and, 1–3
  in North America, 2
  role of, for advance knowledge transfer, 12–16
Forest landscape ecology, 1–2, 6–7, 9–17, 20–21
  collaborative processes, 70–71
  education, 70–71
  knowledge and their interlinkages, 9
  principles, 90–91
  spatial data and tools, 70–71
Forest landscape ecology and spatial scales, 31
  commonsense aspects of, 32–33
  discussion of ecological scale, 32
  importance of, 32
  issues of ecological scale, 35
  observations at multiple scales, 33–34
  principles of landscape ecology, definition of, 34
  synthesis of, 35
Forest landscape ecology in Ontario, applications of, 132, 136–138
  enabling structures, 138–139
  forest management policies and guides, 133–136
  sociopolitical drivers, 133
Forest management
  directions in Ontario, Canada, levels of, 6
  planning in Ontario, tools and applications developed to support, 137
Forest sustainability, 2, 5–6

G

Geographical information system (GIS), 70–71
  specialists, 14–15
Gila National Forest, in New Mexico, 67
GIS, *see* Geographical information system

GIS-based planning and management, Ontario's implementation of, 139
Greater Flagstaff Area Community Wildfire Protection Plan, 83
Great Lakes Ecological assessment, 163t
Grizzly bear (*Ursus horribilis*), 52

H

Habitat models, 135, 142–143
HARVEST LITE model, 161
HARVEST model, 161
Healthy Forests Restoration Act of 2003, 83, 160, 162
Hemlock (*Tsuga canadensis*), 169

I

Integration, for knowledge transfer, 186
Interior Columbia Basin Ecosystem Management Project (ICBEMP), assessment of, 162–163, 165

J

Jack pine–aspen–oak (*Quercus* spp.) forest, 101
Jack pine–black spruce forest, 101
Jack pine-dominated forest ecosystem, 98, 102
Jack pine (*Pinus banksiana*), 57, 98, 101–102, 108, 169

K

Kaibab Plateau, in Arizona, 67, 68f
Knowledge transfer, 70, 211–213; *see also* Knowledge transfer, fundamentals in; Landscape ecological knowledge transfer
  audience, 13
  benefits to developers, 211
  challenges in multipartner landscape, 125–126
  collaboration during decision making, value and function of, 77–80
  collaborative-iterative approach, 16
  decision maker, 7
  definition of, 3
  demand-driven ("pull") transfer approach, 15–16

Knowledge transfer (*Continued*)
  diversity in experiences of, 206–211
  educating stakeholders on application and utility of analyses, 71
  exercise, with natural resource managers, 108
  factors for, 3–11; *see also* Knowledge transfer, factors influencing
  ForestERA spatial decision-support system, 71–77
  forest landscape ecologists and, 1–3
  forest managers, 7, 10–11, 13
  goal of, 207
  is an active process, 206–207
  land manager, 2, 10–11
  land use planner, 7, 12
  legislator, 7, 12–13, 15
  mechanisms, *see* Knowledge transfer
  necessity of, 206
  practitioner, 9, 11, 13–14
  resource manager, 6–8, 12–14
  role of, 141–144
  role of forest landscape ecologists for advance, 12–16
  stakeholders, 7, 12
  successes, in multipartner landscape, 126–127
  supply-driven ("push") transfer approach, 15
  terms, definitions and examples, 13
  users of, 6–7, 9–16, 140–141
Knowledge transfer, at Border Lakes region of Minnesota and Ontario, 111–112
  increasing scale and impact of partnership, 124–125
  phase four: build institutional support, 123–124
  phase one: initiate the collaborative learning process, 112–113
  phase three: early review of the pilot model, 120–123
  phase two: pilot modeling project, 113–120
Knowledge transfer, categories of factors for effective, 183
  effective program-or activity-design principles, 195–202

  empowering institutional environment, 191–195
  philosophical foundation, 184–190
Knowledge transfer, dimensions in philosophy of effective
  engagement, 184–187
  relationships, 187–190
  scholarship, 190
Knowledge transfer, effective program-or activity-design principles
  credible research-based information, 199–200
  evaluation, 200–202
  learner-centric education, 195–199
Knowledge transfer, empowering institutional environment for
  accountability and evaluation for, 192
  knowledge transfer is a choice, 191
  skilled professionals for, 192–195
  stability of funds and nimbleness, requires both, 191–192
  structural implications for, 192
Knowledge transfer, factors influencing
  barriers to knowledge transfer, 9–11
  generation of research knowledge, 3–5
  knowledge users, 6–7
  potential for applications, 5–6
  technological infrastructure, 7–9
Knowledge transfer, fundamentals in, 181
  challenges to, 202–203
  definition, 184
  effective, 183; *see also* Knowledge transfer, categories of factors for effective
  institutions role, 183
  knowledge transfer professionals role in, 182–183
  organizations, 192
  practitioners role in, 182
  principles for success of, 183
  researchers role in, 182
Knowledge transfer, role of forest landscape ecologists for advance
  play an active role, 15–16
  understand the basics, 12–15
Knowledge transfer, successes and challenges in, 89–92

Index

ForestERA data and tools, use of, 85–86
future efforts, changes for, 88–89
obstacles, overcoming, 86–88
Knowledge transfer in Ontario
  use of collaborative–iterative mode of, 152
  use of demand-driven mode of, 152
  use of supply-driven mode of, 152
Knowledge transfer mechanisms
  collaborative planning processes transfer, 82–84
  forest landscape ecological tools during development, assessment, 81–82
  learning process using ForestERA products, supporting, 84–85
  stakeholder needs, delivering data and tools to meet, 81
  stakeholders in project development, engagement of, 80–81
Knowledge transfer principles application, case study
  audience, 145–146
  description, 144
  designing, 145–147
  developing applications, 148
  disseminating findings, 148
  implementing, 147
  sharing results, 147–148
Knowledge users; *see also* Forest landscape ecological knowledge users
  forest resource managers, 6–7
  land-use planners, 7
  legislators, 7
  policymakers, 7
  stakeholders, 7

L

Land and Resource Management Plan planning group, 53, 55, 58
LANDIS forest dynamics simulation software, 99, 115–119, 127
LANDIS model, 50–52
  forest harvest module of, 51
LANDIS modeling project, 119
  parameterization phase, 119
Land managers, 44, 52, 57
  and researchers, 45–46
Landscape ecological knowledge in Border Lakes region of Minnesota and Ontario, transferring, 98–99
Landscape ecological knowledge transfer, 148
  challenges, 149–150
  general lessons for landscape ecologists, 150–152
Landscape ecological knowledge transfer, challenges to
  information overload, 149
  unfamiliarity with landscape ecology, 149
  unrealistic expectations, 149
  viewing GIS technology as a substitute for landscape ecological knowledge, 149
Landscape ecological knowledge transfer, problems in
  audience complexity and diversity, 150
  continuous shifts in organizational priorities, 150
  narrow windows of opportunity for knowledge transfer, 150
Landscape ecological principles, development and transfer of spatial tools, 66–67
Landscape ecological simulation models, points to promote, 35
  comprehensive review and synthesis of scale in ecology, 39
  interactive knowledge transfer, 38–39
  limitations of, clarification of, 37–38
  simplicity in development of, 36–37
  user and use of, 36
Landscape ecological view, adoption of, 6
Landscape ecology, in U.S. National Forests management, approaches for application of science of, 158
  analyses of landscape and regional change, 168–171
  emulating natural disturbance, 173–174
  integrated landscape management, 172–173
  managing roads, 174–175
  National Forest planning, 158–162
  regional and national resource assessments, 162–168

Landscape ecology in forest management, 21
Landscape ecology knowledge, data sources, 7
Landscape ecology researchers and forest managers, differences between, 10
Landscape-level management, 44
Landscape management, integrated, 172–173
Leafy spurge (*Euphorbia*), 72
LEAP, 136–137
Learner-centric education, steps in
  adapt to independent, self-directed nature of adult learners, 198–199
  adopt a minimalist philosophy, 199
  create positive learning environment, 197–198
  document personal and group achievements, 199
  identify audience needs, 196–197
  incorporating range of teaching modalities to accommodate different learning styles, 198
Loblolly pine (*Pinus taeda*), 170
Local resource experts, 46
Long leaf pine (*Pinus palustris*), 167

M

MAGIS model, 161
Malta star-thistle (*Centaurea melitensis*), 72
Management
  planners, 46
  planning, 5–6, 11, 109, 133–137, 142–143, 145, 153, 161
Maple (*Acer* spp.), 169
Marbled murrelet (*Brachyramphus marmoratus*), 159
Matrix-forming forests, 102
Mexican spotted owl (*Strix occidentalis lucida*), 72–73, 75–78, 88
Minnesota Forest Resources Council, 172–173
MLM, *see* Morice Landscape Model
Models, 11, 13, 16, 21–40
Model-development process, 48
Modeling technology to be transferred
  LANDIS, 50–52
  SELES, 48–50

Moose (*Alces alces*), 133, 135, 137
Morice Land and resource management plan, 52
  indicator models, 55
  Morice Landscape Model (MLM), 52–54
  obstacles, 56
  outcomes, 56
  process models, 54–55
  wildland-urban interface and, managing fire risk in, comparison of, 60–61
Morice Landscape Model (MLM), 52–53
  conceptual design of, 54
Mountain goat (*Oreamnos americanus*), 52
Mountain pine beetle (*Dendroctonus ponderosae*), 55

N

National Environmental Policy Act of 1969, 79
National Forest Management Act of 1976, 79, 158
National Forest planning, 158–161
  challenges, 161–162
  concepts from landscape ecology to, 159
Natural disturbance, emulation of, 173–174
NDPEG tool, 136–137
Neutrality, for knowledge transfer, 185–186
Nicolet National Forest, fire risk in, 48, 56–57
Northern Forest Lands Study, 163t, 165
Northern goshawk (*Accipiter gentilis*), 52
Northern spotted owl (*Strix occidentalis caurina*), 159, 162
Northwest Forest Plan, 159
  assessment of, 162, 163t, 165–166

O

OMNR, *see* Ontario Ministry of Natural Resources
Ontario Ministry of Natural Resources (OMNR), 130
Ontario's forest policy framework
  legislative and regulatory requirements, 132
  provincial policies, 132
  strategic directions, 132

Index **221**

Ontario Wildlife Habitat Assessment Model (OWHAM), 142–143
OWHAM, *see* Ontario Wildlife Habitat Assessment Model
Ozark–Ouachita Highlands Assessment, 163t

P

Paper birch (*Betula papyrifera*), 101–102, 169
Parsimony principle, 36–37
Partnerships, 189–190
Passerine birds, 72, 75
Pileated woodpecker (*Dryocopus pileatus*), 135, 137, 143
Pine marten (*Martes americana*), 133, 135, 137, 143
Policies, 24–25, 27, 130–134, 136, 138, 140–142, 144–145, 147–148, 151–152, 182, 210–212
Ponderosa pine (*Pinus ponderosa*), 67–69
Practitioners, role in development and application of knowledge, 182
Procedural language in SELES, 48
Process models
  forest growth, 54–55
  harvesting model, 55
  natural disturbance model, 55
  road access, 55
Professionals, knowledge transfer, 182–183
  skilled, *see* Skilled knowledge transfer professionals
Program, definition, 184
Pronghorn antelope (*Antilocapra americana*), 73, 75, 78
"Pull" transfer approach, 24
"Push" transfer approach, 24

Q

Quaking aspen (*Populus tremuloides*), 101–102, 169
Quetico Provincial Park, 107

R

Red alder trees (*Alnus rubra*), 195
Red pine (*Pinus resinosa*), 57, 101, 119–121, 169

Red-shouldered hawk (*Buteo lineatus*), 133, 137
Relationships, characterization of nature of
  communities, 187
  other institutions, 188
  stakeholders, 188
  teams, 188
Relationships, for knowledge transfer, 187
  collaborative relationship, 189
  cooperative relationship, 188–189
  partnerships, 189–190
Researchers
  and land managers, 45–46
  role in development and application of knowledge, 182
Resource assessments, regional and national, 162–168
  United States General Accounting Office (GAO 2000) suggestions for conducting, 168t
Resource partnerships, for knowledge transfer, 187
Respect for audience, for knowledge transfer, 185
Responsiveness, for knowledge transfer, 185
Rim Country Community Wildfire Protection Plan (GCA 2004), 83
Roads in landscapes, management of, 174–175

S

Scale, 6–7, 10–11, 15, 21–22, 26–40
Scotch thistle (*Onopordum acanthium*), 72
SELES, *see* Spatially Explicit Landscape Event Simulator
Sierra Nevada Ecosystems Project (SNEP), 163t
SIMPPLLE model, *see* Simulating Patterns and Processes at Landscape Scales model
Simulating Patterns and Processes at Landscape Scales (SIMPPLLE) model, 160–161
Simulation models, *see* Forest landscape ecological simulation models

Skilled knowledge transfer professionals, characteristics, 195
 appropriate knowledge of a discipline, 193–194
 awareness of the roles within the knowledge transfer system, 193
 exceptional interpersonal communications skills, 194
 knowledge of instructional-design tools, 194
 personal character traits that enhance knowledge transfer, 194–195
Sociopolitical drivers, forest landscape ecology in Ontario
 global drivers, 133
 local sociopolitical drivers, 133
Southern Forest Resources Assessment, 163t, 166–167
Spatial heterogeneity, 4
Spatially Explicit Landscape Event Simulator (SELES), 48–50
 features of, 50
 "model-building factory" in, 50
Spotted owl
 Mexican (*Strix occidentalis lucida*), 73, 75–77, 88
 northern (*Strix occidentalis caurina*), 159, 162
Spruce beetle (*Dendroctonus rufipennis*), 55
Spruce budworm (*Choristoneura fumiferana*), 101
Stakeholders, in ForestERA project, 70, 90–92
 data and tools to meet needs, delivering, 81
 educating, on application and utility of analyses, 71
 needs assessment, 81–82
 project development, engagement of, 80–81
STELLA software tool, 60

T
Tassel-eared squirrel (*Sciurus aberti*), 72, 76, 78
Technology transfer, collaborative, iterative approach to, 46
 "community of practice" in, 46
Technology transfer, definition, 3, 184
Technology transfer process, 44
 case studies, comparison of, 60–61
 case study 1: Morice Land and resource management plan, 52–56
 case study 2: managing fire risk in wildland–urban interface, 56–60
 conceptual framework of, 45–48
 modeling technology to be transferred, 48–52
Timber, 130, 165–166, 170, 175
 harvesting, 166
 supply, in North America, 5–6
Timber harvest allocation model, *see* HARVEST model
Transfer specialist, 12

U
U.S. National Forests management, approaches for application of landscape ecology to, 158
 analyses of landscape and regional change, 168–171
 emulating natural disturbance, 173–174
 integrated landscape management, 172–173
 managing roads, 174–175
 National Forest planning, 158–162
 regional and national resource assessments, 162–168
USDA Forest Services, 93, 106, 111–116, 158–167, 171–172, 175
 Boundary Waters Canoe, 105
 Coconino National Forest, 82
 Crown lands in Ontario, 107
 recovery plan by, 88
 roads management by, 175
 Rocky Mountain Research Station, 82
 Southwest Regional Office, 85
 Superior National Forest, 98

V
Visualization tools, 172

W
Western balsam bark beetle (*Dryocoetes confusus*), 55

Western Mogollon Plateau adaptive
　　landscape assessment process,
　　73–74, 82–85
　workshops, 92
White pine (*Pinus strobus*), 101, 119–121,
　　169
White spruce (*Picea glauca*), 101, 167
White-tailed deer (*Odocoileus virginianus*),
　　44, 135, 137, 143
Wildfire, 67, 69, 73, 85–86
Wildland–urban interface, managing fire
　　risk in, 56–57

Morice Land and resource management
　　plan and, comparison of,
　　60–61
　obstacles, 59–60
　outcomes, 58–59
Woodland caribou (*Rangifer tarandus
　　caribou*), 52, 133, 135, 137
Wood-pewee (*Contopus sordidulus*), 76

Y

Yellow star thistle (*Centaurea solstitialis*),
　　72

Printed in the United States of America.